MW00812122

Floating to Space

The Airship to Orbit Program

All rights reserved under article two of the Berne Copyright Convention (1971).
We acknowledge the financial support of the Government of Canada through the Book Publishing Industry Development Program for our publishing activities.

Published by Apogee Books, Box 62034, Burlington,
Ontario, Canada, L7R 4K2, http://www.apogeebooks.com
Tel: 905 637 5737

Printed and bound in Canada
Floating to Space - The Airship to Orbit Program by John Powell
©2008 Apogee Books
ISBN 978-1-894959-73-5

Floating to Space

The Airship to Orbit Program

John Powell

An Apogee Books Publication

Contents

Dedication

To my son Aubrey: May his future be filled with cities in the sky.

This book is also dedicated to all those out there attempting out-of-the-box, crazy and unaccepted ideas.

They persevere, even through ridicule and impossible odds. Even though most will fail, they are the ones that our future will be built upon.

Acknowledgments

To all the JPA Team. They are the ones making it happen.

A big thank you to Natalie Vollrath for reviewing my words and making them better.

JP

Chapter 1
Where are all the spaceships?

We have finally made it to the twenty-first century. We are living in the future humanity has dreamed about. Many of the wonders promised for this era are actually here. Cell phones, the Internet, genetic engineering, and personal computers were all the stuff of science fiction just a few years back. The future is here. But wait, something is missing. There is a glaring hole in our attainment of the future. Where is the Starship Enterprise? Where's the big 2001 A Space Odyssey space station? Where are you, Captain Kirk? Where are all the spaceships?

Thirty years ago, the world was poised at the edge of the solar system. In their race to the moon, the US and USSR solved all the major problems of space flight. All the elements of space flight were in use - powerful rockets, navigation, life support systems, and the will to go. We went, we flew, we landed and we even walked around on an alien world. Then we stopped.

What happened? Why didn't we carry through? The problem was not the technology, the problem was the goal. The goal we had left us without the will to continue on. The planet was embroiled in a cold war contest for dominance. The goal was not to send man out to the stars; the goal was political one-upmanship in which the stakes were the world itself. With the stakes this high, money was no object. Both sides poured all of their financial, intellectual, and political resources into the space race. Reaching to space was more than the science or new frontiers; it was a measure of the success of political systems. War and disaster are usually required for this level of effort and only then when backs are against the wall. The results were spectacular. Both countries achieved stunningly successful programs that are impossible to repeat. Unless some stray asteroid heads for a collision course with the Earth or the aliens invade, we won't see that intense level of commitment from national governments for space again.

We live in the peculiar aftermath of that space race. We have the technology; hundreds of people have already been to space. Lots of very smart people out there are putting a lot of effort into space, but still, where are all the spaceships? Progress has been so slow that space travel is a history lesson instead of a science lesson in schools. The road traveled before isn't getting us there. The old joke hits you right in the face, "You can't get there from here."

The fact is, there are only a few makers of spaceships in the world; By spaceship, I mean a craft people can ride in. Of the top three; the Russian Space Agency is broke, NASA has not designed a new spaceship in twenty-five years and the Chinese are flying upgraded Russian copies.

What is a self-respecting space wannabe humanity to do? The Columbia tragedy has left the United States space program in disarray. The return to flight of the Space Shuttle Discovery has only served to highlight the flaws in the system. The vehicle not only needs extensive repairs between flight, but now it also needs repairs before it can even return from space. The President's new mandate to return to the Moon and then go to Mars calls for a new spaceship, the Crew Expedition Vehicle, CEV. The goals of this vehicle have fallen short even before the real vehicle even exists. Current NASA estimates call for ten years and ten billion dollars to create a craft that is smaller and less capable than the existing space shuttle. The Russian space program is hampered by huge economic problems. The Chinese are now joining the space community; however they are decades behind. All these efforts are only improvements on the existing technology, large rockets fired directly into space. The focus is on lighter materials, more efficient engines, and better electronics. This is fine-tuning. Fine-tuning does not create dramatic change.

These improvements do little to aid in humanity's reach into space. If governments are unwilling or unable to make the big leaps, can anyone else? There is hope coming from the commercial world. Private companies and organizations are charging ahead. Scaled Composites in California has flown their own astronauts on suborbital flights. They broke the myth that only governments can send people to space. Armadillo Aerospace in Texas is developing a vertical takeoff rocket reminiscent of the NASA DC-X program. Another company in California, SpaceX, is developing a new rocket to launch satellites. Their rocket, the Falcon, may result in the first truly commercially viable satellite launch system.

Without the strings and confinements of a government program, they are free to experiment and try new methods and technologies. Many other small companies are trying to refine the rocket. The great successes and great failures of these groups are putting excitement back into space.

The achievements these groups have made for amazingly few dollars put NASA and the other government space agencies to shame. In a few years, these new rockets will rank right up there with the government space programs, launching rockets with and without crews. The problem, however, remains. It's not enough. The new generation of private rockets (I put space planes that act like rockets on the way up in the same category), will be more efficient and cheaper; however, it still is only fine-tuning what is already there. These new vehicles are like a new fuel efficient economy car. It's superior to last year's model, but it still has four wheels, runs on gas and only goes where the roads are. These new spaceships will not fundamentally change humanity into a spacefaring society.

What is the fundamental issue? It is hard and expensive to fly crewed vehicles to space using current methods. It is an obvious but often-overlooked fact that if humans are going to expand into space, you must first get there. The difficulty of this initial step into space, flight to Earth orbit, that has held us back. Current space flight technology is too expensive, too dangerous, takes an army to get ready, carries too little, and will never open up the space frontier.

The Wan-Hu Method.

One of the earliest stories about rocket travel is that of Wan-Hu from China. Although consider apocryphal, the story does serve to make a point. Like most of the people running space programs today, Wan-Hu was a government official. His spaceship was a chair with two kites mounted on top and forty-seven rockets mounted on the back. The configuration is still seen today in the Space Shuttle. The story doesn't mention a countdown, but it does describe a large ground crew. Forty-seven members of his "mission control team" rushed forward, each with a flaming torch and lit the rockets. With a loud boom, Wan-Hu and his starship chair were gone. Now unless we find his chair on the moon someday, we must conclude that this was a bad idea - a very bad idea.

Yet, this is exactly how humans travel to space today. When Yuri Gagarin became the first human in space, he used the Wan-Hu method. When John Glenn orbited the Earth, he used the Wan-Hu method, and when Neil Armstrong stepped on the moon, he got there using ... you guessed it...the Wan-Hu method. The Space Shuttle, the Russian Soyuz, the French Ariane, the Chinese Long March rocket and all the rest are just bigger, more expensive versions of Wan-Hu's chair. The method of sitting on the biggest controlled explosion you can make continues to be the technique of choice for space travel.

The question we must ask is, is there another way?

For over a hundred years, there has been a cheap, safe, and fast technology that has carried people and equipment to the edge of space. Without using rockets, this technology carries tons of cargo to the top of the atmosphere every year. That technology is balloons. Balloons!?!? What do balloons have to do with flying to space? Everyday balloons fly to the edge of space. Before the existence of NASA, the United States flew manned flights to the edge of space with crews in spacesuits. Today the top of the atmosphere has become the new "garage" for hundreds of amateur scientists and experimenters. With new available materials and techniques, the performance and abilities of balloons are dramatically increasing. It's time to fly them over the edge. It is time to push this age old technology the last mile. It is now within our ability to essentially float to space.

Slingshot and Space Policy, a new story.

A thousand years ago in the old country, (pick your favorite), there was a village along a wide river. The village was a center of commerce. Their main focus in life was trade. They had one problem...a natural barrier limited their trade. Only the villages on their side of the river were accessible.

For many years they looked with wonder at the other side. " What new traders exist over the waters?" was their common thought. One day, one of the village elders decided to stop talking and find out what was on the other side. Many ways were proposed for getting across the river. The wise men and the shamans all huddled together and made their plans. After much deliberation they decided to go with what they knew. They would build a mighty slingshot.

The slingshot had been invented years ago to hurl rocks at the next village. During the wars, the slingshots were all that stood between the villagers and certain death. The best and brightest minds in the village were given all the gold they required to build the ultimate rubber slingers of death. This was many years ago and, for the most part, all had become peaceful with the neighboring villages. However, the slingshots still stood guard. So when it came time for the first leaps across the river the villages used the giant slingshots from the wars. As the slinging progressed, the old war rubber bands were improved and enlarged. Eventually, a person could be flung across the river. These people were the Sling-nauts.

Sling-nauts were the heroes of the village. The sling-nauts were highly skilled professionals. It took years of training to hurl though the air and land just right. Slinging also took a great deal of courage. Sometimes a Sling-naut would be killed, but that is, sometimes, the lot of a hero.

The day of the slinging was always a great day of celebration, and everyone in the village would come down to the river to watch. One morning before a Sling a young boy named Aubrey wakes up early and goes to watch. It is still early and the crowds have not yet arrived. Walking close to the shore, Aubrey comes across an old fisherman tending to his boat. Not knowing any better, Aubrey asks him, "why don't you use your boat to cross the river?"

The fisherman laughs, 'Kids...,' he muses to himself. The fisherman explains that boats go almost to the other side all the time. Every few years some intrepid fisherman skirts the far shoreline, but a boat hasn't been built that can make it the last leg to the other shore. Of course the fisherman also knew that a boat could never be used to cross all the way to the other side, what nonsense.

Aubrey, being only eight years old asks the perpetual child's question, "Why?, Why can't they go all the way to the other side?"

The old fisherman just smiles "Who ever heard of a boat that can go all the way to the other side?"

Wandering further down the shore, Aubrey finds the slingshot team busy with checklists and procedures. One of the team is wearing a white vest and a crown of Eagle feather, Aubrey guesses that

he's the boss and goes to ask him his question. "Why can't boats be used to get to the other side?".
The chief keeper of the slingshot then explains all the reasons why boats won't work.

"Boats?, You couldn't possibly use a boat to cross the river. It would be washed downstream.
A boat would be too slow. A boat large enough could never be built. Boats have been around for a long
time, if you could cross a river by boat someone would have done it by now".

The hard truth is that it just never occurred to them to take a boat all the way across.

The current stable of rockets used to launch satellites today are the nuclear missiles of a generation ago. They are our slingshots. They were designed for an extremely precise mission where cost
was no object. This is not a negative indictment. When you are carrying atomic bombs, only the best
will do. However, they are ill-suited for the bulk cargo-runs to space that are needed by commerce
today.

More than just a new slingshot is required. What is needed is a whole new set of roads. Is
there a way to "cross the river" without a slingshot? Is there a way to reach space without rockets?
Could it be that the way is right in front of us? Is the way right in front of our eyes, but it just hadn't
occurred to us yet?

Balloons are our boats. Manned balloon flight has been with us for well over two hundred
years. They have carried people and large multi-ton experiments to the edge of space. The last mile to
the other shore is not a trivial one. Many challenges exist to be overcome. Yet, there is now a way that
balloons can go that final distance.

Balloons, or more accurately, airships can now be built that can reach space. But wait, balloons fly in the atmosphere. The air surely stops well before you get to space, right?

The common notion is that of our atmosphere being this thin, surface-hugging sheen, just
above the planet. In fact the Earth's atmosphere extends thousands of miles into space. When the Space
Shuttle flies, the astronauts can see a glow from the tail. This glow is caused from the tail hitting air
molecules at orbital velocities. It was the air all the way up in orbit that brought down both the US
Skylab space station and the Russian Mir space station. The atmosphere may be thin up at those
extreme altitudes, yet it can still have powerful effects. It's this power in the upper atmosphere that is
the key. For the current way of reaching space, the atmosphere is the enemy. For airships flying to space
the atmosphere will be a friend. Now the atmosphere will be the ladder.

This ladder has three rungs. On each rung is a lighter-than-air vehicle.

With our current technological abilities, a single-piece rocket can not reach orbit around the
Earth. The rocket needs to be broken up into steps, or stages. Most rockets use three to four stages to
reach the speed necessary for orbit. Airship technology faces the same problem. Flying an airship
directly from the ground to orbit is not currently possible. An airship large enough to reach orbit would
not survive the winds near the surface of the Earth. Even a five mph wind would create hundreds of
tons of force on its large sides. The vehicle would not survive taxiing out of the hanger, let alone the
turbulent dense air of the lower atmosphere. Conversely, an airship that was strong enough to fly from
the ground to the upper atmosphere would not be light enough to reach space. Just as with rockets, airships will overcome this problem by the use of steps or stages. Each step can be optimized to the environment it will operate in. There are two very different operation environments, the lower and upper
atmosphere. Unlike rockets the lower stages can't carry the upper stages on its back. In addition there
needs to be a way to interface between the two. The resulting configuration is a three-part architecture
for using lighter-than-air vehicles at each step to reach space. The name of this method of reaching
space is called "Airship to Orbit" or ATO.

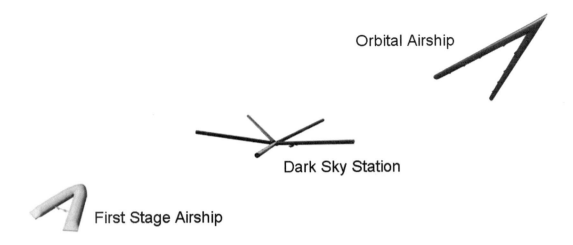

Figure 1-1: Three part architecture for reaching space.

The first part is a pure atmospheric airship, a high flying Goodyear blimp. This airship will travel from the surface of the Earth to 140,000 feet. The vehicle is operated by a small crew and can be configured for cargo or passengers. Propellers designed to operate in near vacuum of the upper atmosphere drive the vehicle.

The second part of the architecture is not an aircraft, but a station. This is a permanent, crewed facility parked at 140,000 feet. These facilities, called Dark Sky Stations (DSS), act as the way stations to space. The DSS is the destination of the atmospheric airship and the departure port to orbit.

The third part of the architecture is an airship/dynamic vehicle that flies from the DSS to orbit. In order to utilize the few molecules of atmosphere at extreme altitudes, this ship needs to be big. The initial test vehicle will be 6,000 feet, (over a mile) long. The airship uses buoyancy to climb to over 200,000 feet. From there it uses electric propulsion to slowly accelerate. As it accelerates, it climbs dynamically. Over several days it reaches orbit. Once it is in orbit the airship, now a spacecraft, can deploy satellites, deliver personnel and cargo to space stations or head to more exotic locations.

The Atmospheric airship

The first step on the road to space is just getting off the ground. Airships have traditionally been low altitude vehicles. A modern advertising blimp has a maximum altitude of only ten thousand feet. Although in order to get good exposure for the big advertisement on the side, they typically fly at altitudes of less then two thousand feet. The highest flying production airships were the World War I dirigibles the Germans used to drop bombs on England. These airships were known to fly up to 20,000 feet to escape being shot down by enemy aircraft. Until recently no airship has gone higher. The first stage airship for ATO will need to reach 140,000 feet, higher than any airship has flown before. It will require materials and techniques not used by airships before. This vehicle must be robust enough to withstand the weather of the lower atmosphere- wind, rain, and lighting. This toughness translates into weight. Weight limits the altitude that the vehicle can reach.

Figure 1-2: The First Stage Airship lifting off.

The key to reaching these heights are the materials. High strength, lightweight materials are now available. Carbon fiber, rip-stop polyethylene's, and ultra-thin films are the materials of choice. These materials will be used to create the gossamer structure needed to fly to extreme altitudes.

The other critical technology needed for the high altitude airship is the propeller. For a propeller to work in the near vacuum at the edge of space, it needs to be extremely efficient. The propeller needs to impart energy from the motor into each and every air molecule that happens by. New propeller designs have been tested at the edge of space do just that.

JP Aerospace has developed a high altitude airship specifically as the first stage of ATO. Several versions of the airship, called the Ascender have flown. All of the critical elements such as the propellers have been flown and tested in the upper atmosphere.

Port at the edge of space

The idea is simple. Seaports can be found at the waters edge. Spaceports should be at the edge of space. The obvious question is "What do you set it on?". Seaports have this thing called a shore to pour the concrete on. There is something at the edge of space that can be equally substantial-hundreds of millions of tons of air. There is nothing insubstantial about the Earth's atmosphere. The idea of floating piers and docks are millennia old. This dock just floats a little higher. 140,000 feet is the launching point for space. This is where the spaceport will be.

This spaceport will be a floating structure two miles in diameter. It will house a crew of twenty to thirty. Not quite a city it will at least be a village twenty-eight miles up. The current design looks like a giant starfish. Five large cylindrical arms hold hydrogen providing the lift. The arms are joined at a central hub.

Figure 1-3. Dark Sky Station.

This high altitude outpost will serve several functions. It will be a transition area, the destination of the airships leaving the Earth and the disembarking point of giant gossamer vehicles needed to ascend to orbit. It is the construction facility for the large orbital vehicles. The station is also an excellent venue for near-space tourism.

Several small prototype Dark Sky Stations have flown. Watching a Dark Sky Station test flight is an amazing experience. DSS test vehicles have ranged from twenty-seven to fifty-seven feet in diameter. The incredible moment comes when the huge vehicle leaps off the ground and rushes skyward, yet doesn't make a sound. It's the perfect metaphor of this new direction in space travel.

Airship to Spaceship

To use the few molecules that exist between 140,000 feet and space, the orbital airship must be both light and large. To reach the velocities needed for orbit, the craft needs abilities of both an airship and an airplane. For the first part of the trip, the orbital airship uses the same method as conventional airship; it floats. From 140,000 feet to 200,000 feet, the hydrogen inside its lift cell provides the buoyancy. From 200,000 to 260,000 feet is the transition range. At the bottom of the range, the ship is still mostly an airship getting its lift from the hydrogen inside. The vehicle is beginning to get some lift from the wings. As it climbs higher and higher, lift from the hydrogen diminishes and the wings carry more of the load. Above 260,000 feet the vehicle still uses the atmosphere, but now acts like an airplane. It will use lift from its wing-shape to carry it the rest of the way. This need to transition from airship to airplane dictates the shape of the craft. Existing airships do generate lift. The Goodyear blimp is actually heavier then air. It uses the lift generated from its body to fly upward. If the engines were to fail, the blimp would slowly drift to the ground. The traditional cigar shape is ideal for the low speed flight. The Graf Zeppelin, a very fast airship, could travel at eighty mph. The orbital airship will need to travel at 24,000 mph. To create lift at these velocities, the blimp shape is too inefficient, as it creates too much drag. To create lift efficiently, the airship needs to be shaped more like an airplane. However, here lies a dilemma. Thin wings that are good for lift and low drag are not very good for holding large volumes of hydrogen for buoyancy. The solution is to think big. An orbital airship with a low drag shape will need to be over a mile long to have the necessary volume. To meet the requirements of stability and low drag in the low-pressure, high-speed flight envelope, a large narrow swept wing shape is used. The resulting craft looks like a giant 'V'.

The vehicle is a contradiction, a light gossamer structure flying at twenty-four times the speed of sound. Propulsion comes from low thrust electric rocket engines with the power provided by solar panels and fuel cells.

After completing its mission, the orbital airship will use air friction, drag, to slow down and reenter the lower atmosphere. Like watching the climb to orbit backwards, the reentry will also be a slow process spread over several days. The ship will then return to the Dark Sky Station. The orbital Ascender will live its entire life in the sky. Built in the upper atmosphere and operating between 140,000 feet and orbit, the ship will never touch the ground.

Miniature, twenty-foot long versions of the orbital Ascender exist now in the lab. Soon they will be deployed at 120,000 for high speed glide tests. Called Mach Gliders, these mini x-planes will be the tools for gathering data needed for the ATO program.

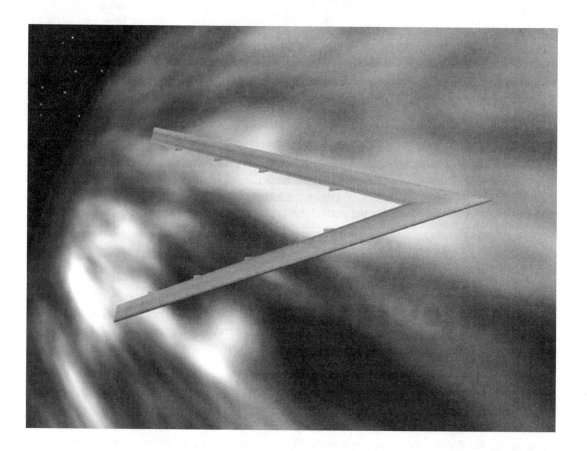

Figure 1-4. Orbital Airship

What of the road itself?

We typically think of us being down here and space being "out there", however, little thought is given to what's in between. What we're interested in is that area of space below Earth orbit. It has been given an array of names and descriptions—the upper atmosphere, suborbital space, near space and even the ignorosphere, for most ignored part of our planet. This is the region between eighteen and sixty-two miles above the Earth. It sits above where airplanes fly and below where spacecraft circle. Once thought of as empty, it is now known to be the home of an amazing array of phenomenon. In addition to providing a ladder to the stars, this region holds its own secrets waiting to be discovered.

Life, Sprites, the blue flash and clouds in space.

There is nothing more familiar in the sky than a cloud, but clouds in space? Many things in space are referred to as clouds, such as large groups of stars, remains of stellar explosions, dust lanes the size of galaxies and similar grand phenomenon. However, these are only metaphors for clouds; they are not the puffy water vapor and ice-crystals-in-the-sky variety of object we associate with the word "cloud". There are real honest-to-goodness clouds made out of water that form fifty miles up in honest-to-goodness outer space. They have many names Mesospheric Ice Clouds, Noctilucent clouds, and even "night-shining" clouds. It is likely that you have seen them without even realizing it.

Life at the Edge of Space

It's hard to image the clear seemingly empty region of near space being a home for life. It's even harder to picture near space teeming with life, yet the current data suggests just that. The upper reaches of the sky may be the largest living environment of the Earth.

In January 21, 2001 the India space program in partnership with Cardiff University, in the United Kingdom, launched a balloon carrying a very special sampling device. The device was a cryosampler manifold with stainless steel probes. It was designed to take samples from the stratosphere under extreme sterilized and aseptic conditions. They were searching for life twenty-five miles up. At altitude the chambers were opened and the atmosphere was passed through micropore filters. The samples showed evidence of living cells in abundance. A scanning electron microscope was used to closely inspect the samples. The photographs taken clearly show microorganisms.

This confirmed early 60s NASA balloon experiments. When originally conducted these tests showing life in the upper atmosphere were considered to be so outrageous that the data was dismissed and then forgotten. There is evidence that life can be found even higher up. When you see a falling star streaking across the sky you don't usually find your self thinking, "Wow look at that life up there!" However scientists in England have evidence that, in part, life is exactly what you're seeing.

Spectral examination was conducted of the Leonid Meteor shower. Researchers then heated bacterial samples to the temperature of the meteor trail. The spectra of the meteor trail matched that of the heater sample of e.coli bacteria. They concluded that "the meteor trail fingerprint is most likely due to a population of bacterial particles in the mesosphere that have been transiently heated to temperatures approximately 400 degrees Kelvin".

'Giant red plasma volcanoes in the sky' sounds like the title of a bad science fiction movie. The no less colorful description given to these objects is 'Sprites'. Sprites are just one of a whole family of large electrical phenomenon that includes: Elves, Gnomes, the Blue Flash and Tendrils. Don't let the whimsical names fool you. These are electrical discharges that make the biggest lighting strike look like a spark. Very little is known about Sprites and the rest of this atmospheric menagerie. Their impact on telecommunication, reentering spacecraft and weather may be significant.

Ten years ago we knew and could prove that water ice could not exist fifty miles above the Earth. Now, we can see Mesospheric Ice Clouds with our own eyes, if we know where to look. What else could we discover in the ignorosphere if we could actually go there? Could our first discovery of life in space be right above our heads?

From life at the edge of space to mile tall plasma volcanoes the exploration of this region is more akin to investigating another planet than anything on Earth. ATO technology can provide the means to explore this new realm.

Lower Cost

Low cost, routine access to orbit has been the Holy Grail of the space community. Like the quest for the grail, many have pursued it. These grail-seekers have made a few good moves and a few bad ones. The quest is a noble one; for just like the power of the legendary grail, once a cheap way to get to space is found, all things are possible.

Every bit of equipment put into space is made as light as possible. This is directly due to the extreme cost of getting each pound there. This process makes each item going into space an engineering marvel and as a result, it costs a fortune. The high launch cost also drives up the required reliability for any part; whether it is a communication system or a nut. If it costs forty million dollars to launch your satellite, the satellite must be worth the launch cost.

What type of things do you spend forty million dollars to ship? The easy answer is "nothing that you could buy at Walmart". Think about what would happen if it cost a company forty million dollars to ship your car across the ocean. The first thing is that people would stop shipping cheap cars. No one would pay forty million in shipping for a subcompact econo-box car. The car would need to be built with absolute perfection. Forty million dollar shipping costs mean you couldn't afford to ship another if the first one goes wrong. Thankfully large cargo ships are cheaper to operate than rockets. Cargo ships are more expensive to build, but they have the benefit that you don't throw them away after each delivery. Floods of low cost products, including very cheap cars are the direct result of low-cost transportation. The same effect will take place for space products. Right now there is no such thing as a cheap discount satellite. Even the thought of a discount satellite is considered outlandish. The cost of launch has had a powerful effect on the thinking of the industry. Low cost space transportation opens the door for low cost products to enter. The Yugos and Volkswagens of space will have their place alongside the Mercedes.

Airship to Orbit can dramatically bring down the cost of getting to space compared to traditional launch systems. Four areas sum up why the costs are lower—full reusability, lower risk, bulk cargo and lower energy density.

Our first "reusable" spaceship, the Space Shuttle, was a grand experiment. With it we learned what space reusability really means. Unfortunately the shuttle fell far short of real reusability. After each flight the Space Shuttles main engines need to be removed and overhauled. That big orange external tank burns up on each flight and needs to be replaced. The big side booster rockets just about need a complete rebuild after each flight. Reusability needs to be refueling, cleaning up the trash, inspecting the vehicle before taking off again and that's it. We know how to make sophisticated reusable vehicles. The next time you are at the airport, see if you can spot one. Anyone who thinks a 747 airliner is not as sophisticated as a spacecraft has not seen the inside of the cockpit.

Safer Climb to Space

To reach orbit, a spacecraft need to be moving at 17,000 mph. Conventional launch vehicles are dragsters; they get up and moving to the orbital speed in a hurry.

Riding the Space Shuttle or any other rocket is like riding an explosion. The crew of the spacecraft sits on top of thousands of tons of the most combustible fuel on the planet then some fool lights the whole thing on fire. Millions of pounds of explosive thrust then hurls the spacecraft upward, crushing the occupants and the structure with unbelievable force. After a few minutes of extreme vibration, detonation and down right violence, the ship and crew are in orbit. At mission control there is usually a cheer. Ask yourself, if every time a plane took off without exploding and everyone at the airport gave a cheer and a sigh of relief, would you get onboard?

With ATO you still need to accelerate to the velocity of Mach 24 to reach orbit. The difference is time. The acceleration and the energy to produce the acceleration is spread out.

There is safety in time. By traveling to space in days instead of minutes, there is no sudden release of energy. Without a need for a sudden release of force there is no need for volatile fuels. Low thrust electrically powered engines will do the job.

One of the major costs of reaching space is insurance. A safer method to get to orbit will directly decrease insurance costs.

Airships have a single scene that has been scaring people and creating a bad image for over seventy years. That scene is that of the Hindenburg burning and crashing to the ground. In spite of the fact that scientists and engineers have shown that the lift gas, hydrogen, played no role in the fire and that most of the passengers survived, that one moment has characterized airships as dangerous flying bombs ever since. The reality is that modern airships and blimps have an incredible safety record. In most airships non-flammable Helium is used as the lifting gas. In the Airship to Orbit architecture helium is used for the lower atmosphere airship eliminating any possibility of explosion. Hydrogen is used on the high altitude station and on the orbital airship, where there is too little oxygen to react, eliminating the possibility of explosion.

When problems occur in a conventional rocket there are only seconds or even fractions of seconds before action must be taken. With ATO, fractions of a second stretch out to hours.

Example: Shuttle main engine failure. There are only moments to separate the shuttle orbiter from the rest of the components. It is estimated that the chances of survival for the crew in this situation are thin. If the main engines fail on an airship heading to orbit, you have a meeting. Maybe you want to talk to experts on the ground. Perhaps the technicians could stroll over to the engines and see what's wrong. There's no hurry. During this time, the airship slowly begins to drift down with no danger to the crew, the vehicle, or the payload. If the problem cannot be corrected, the airship simply glides back down to the floating station.

Safer reentry

Capsules blaze like fireballs, old space stations burn up, and seven American astronauts tragically died on the space shuttle Columbia. Coming back has always been the most dangerous part of space travel. It's enough to make you think that the Earth is surrounded by a ring of fire. What is really happening here? Friction is happening. Reentry is the ultimate rug burn. Spacecraft slam into the atmosphere at 17,000 mph and all those air molecules do their best to melt everything they touch. If going fast burns you up, then just say no, and come back slow!

Despite the hazards posed by a launch, it is the return home that has the mission controllers holding their breath.

With the current method of high speed reentry there are no options. Tragic accidents in both the Russian space program with Soyuz and the NASA with Columbia have vividly shown the danger.

The Mir and Skylab space stations didn't need any fuel to reenter and neither will the orbital airship. However, the results will be different. Drag from the atmosphere will bring airships down from orbit. The drag is so significant that orbital airships may need to keep their engines running just to remain in low Earth orbit. The orbital airship will use this drag to start the reentry process. With its ability to generate a lot of drag at orbital altitudes, the orbital airship can bleed off velocity while it's still high. Just like the slow climb up to orbit, it will very descend slowly. By the time the vehicle is low enough in the atmosphere where heat is a problem, velocity will be low enough where heating is not a problem.

Scalable technology

Rocket technology scales poorly. The largest rocket ever, the Saturn-five moon rocket, could only carry two small spacecraft and a crew of three. Airship to Orbit technology can be scaled up with bigger vehicles to accommodate larger cargoes. Increased size, in contrast to rockets, increases safety.

Bulk Access to Space

What could you do with bulk access to space? For manufacturing, space based production of mass-produced goods becomes possible. In the view of manufacturing, reducing launch costs equates to reducing shipping costs. Space entrepreneurs have been searching for decades for the "killer app," the one high value product or service that has a great enough return to make a space business viable. If gold were found floating in orbit, you could not make money by retrieving it.

What product could have a great enough value that manufacturing in space could turn a profit? No one has any idea.

Thousands of tests have been conducted on production in space. Nothing has been found that can make a profit. With low cost flight to orbit, all sorts of products can be made in space.

With bulk access to orbit, space stations and other large structures could be placed into orbit intact. This can greatly reduce or even eliminate the need for on-orbit assembly. Solar power satellites generating electricity to be beamed down to Earth with microwaves becomes practical.

One of the goals of the Space Shuttle was to retrieve satellites from orbit and return them to Earth. The Shuttle was never able to really accomplish this goal because it can't reach where most of the satellites are. An airship in orbit can continue to use its solar/electric propulsion system to go to the higher orbit and retrieve damaged satellites. The satellites could then be refurbished on Earth and sent back up.

In addition to complex equipment and machines, supplies such as rocket fuel could be carried. This allows spacecraft going to the moon and beyond to be launched from the Earth with empty fuel tanks. Launching "dry" reduces launch costs and increases safety. Space transportation can begin to look like the transportation of any other good or service. Cars are not shipped from Japan to the United States with full tanks of gasoline. Neither do you throw the car away when it's out of gas. This is another example of how low launch costs can make everything easier. When a spaceship is out of refuel, refuel it! It all hinges on the cheap shipping of the fuel.

Show Me the Money

Many promising space projects have failed, or worse, never tried because of a lack of funds. Planning for the financing of a project is just as important as how much thrust a rocket has, or how much power a battery produces. The easy method, having the government give up unlimited tax dollars, just isn't available for this new space program. The traditional method of financing projects or businesses, mortgaging the house, borrowing from friends and family cannot generate the large sums needed. Standard investment models are structured for faster rates of return than a space venture can deliver.

Fortunately, opening up this new frontier doesn't need to bankrupt us, nor do we need millions of dollars from venture capitalists. What truly separates ATO from other space projects, from a business prospective, are early customers. Many space projects do not see their first paying client until the rocket has lifted off the pad at least once. This occurs many years into the project. ATO can be a "pay as you go" enterprise. Tourism, sponsorship, and transportation services can pay the way. Passengers on balloons already make up a multi-million dollar a year business. Sponsorship on balloons and blimps make up several more millions. Tourism and business already are a match for lighter-than-air technology. Couple this with a real cargo carrying capacity and the ability to actually go somewhere, and you have a huge business potential ready to be tapped.

Floating cities were once only found in science fiction books and movies. The technology to make them a reality exists now. First with small labs, then on to villages, the progress to cities at the edge of space is already underway. For Airship to Orbit, the floating station is a port to the heavens. It can also be a destination in its own right. A new place to explore, colonize, and expand the vista of mankind.

There is another way.

There is another way to get to space. The oldest aviation technology can be used along with the ignored upper atmosphere to propel people and cargo to space.

Building balloons that scream at twenty-four times the speed of sound and sending them into space will not be easy. A great deal of progress has been made, but a large amount of work is still ahead. Engineering challenges to overcome include developing lightweight power generation, lightweight electric engines, and managing drag at high velocities on extremely large vehicles. The greatest challenge is not engineering, but perception. Come on...supersonic blimps to space? Mile-long aircraft? Like many new ideas, their acceptance travels in stages, from "it's insane", to "well, maybe," to "I knew it all along." The force that will move this concept from one stage to the next with be the development mission and test flight.

Near space is an incredible place. It holds solutions for low cost access to space, and it is the most amazing and completely overlooked environment of Earth. We know nearly nothing about its realm, and yet it's Earth's largest environment by any measure. Life may exist there in near space conditions. Solving this mystery could provide answers to life beyond the Earth.

A great deal is still unknown about the near space environment. Everyday, more is being discovered; however, we need to be there with our instruments and our minds and ourselves to pry open the secrets that are there. How long will it take? How fast could scientific discovery have moved if Lewis and Clark had an electron microscope, a DNA sequencer, and a super computer with them as they explored the West? We have a brand new environment before us. It's time we went there to see what there is to find.

Airships traveling directly to space will completely change the nature of space travel. No longer do you need the "right stuff." This technology will change humanity into a truly spacefaring civilization. Little new is needed; it's just the road not yet traveled.

There is a new world, and it has been right there all along. It's time to float on up and take a look.

It's the twenty-first century. Along with the cell phones, computers and Gameboys, we need some spaceships. We've been timidly wading out into the ocean of space; it's time to send out the fleet.

Chapter 2
Arriving at the Terminal

To give you a feel of what it would be like to go to space on a balloon, we are going take a step into the near future. Our hero on this adventure is Aubrey, the Network Engineer and we will follow his steps as he floats to space.

After arriving at the spaceport prior to departure time, Aubrey now has time to hang around. This doesn't bother him though; at only twenty-six, his life has already become so frantic, he enjoys these tiny respites. "Thank God space travel is a slow process," he thinks. This is a thought that would shock his early astronaut forebears.

Looking out from the terminal, Aubrey involuntarily zips up his jacket. It's warm inside in spite of the 30-below-zero temperature outside. He can hear the Alaskan wind buffeting the big windowpane. With his destination circling high above the Arctic, Aubrey knows that the Kodiak Spaceport in Alaska was the best departure point. Yet, already he missed California, but a job is a job, and there was a need for a network engineer on Mars. Outside the window a huge black shape dominates the runway in the distance.

"'Buck Rogers or the Graf Zeppelin?'" Aubrey ponders. The big ship looks like a cross between a toy rocket and the Goodyear blimp. He had seen pictures of these high altitude airships before. However, not until a 747 taxies across in front of the massive vehicle and looks small does he appreciate just how big the airship is. "Three jumbo-jet-lengths I'd guess," he says out loud. "Aw, that's just the baby," says a dark-haired woman next to him. "First trip topside?"

Aubrey notices the woman is wearing a flight engineer's uniform. She has short cropped brown hair, and her belt is covered with electronic gadgets. " I'll just be passing through, I'm heading out to a lab at Sagan Station on Mars...you're right though, I've never been up top."

"Wait till you see the big guy," the woman says.

"I'm sure we'll all be thrilled," a squeaky voice piped up. "I'm sorry," Aubrey said, with some embarrassment. "It's my iPal. I downloaded a "sarcasm personality" upgrade. I thought it would help with the long trip."

"I'm not used to the scale of these things," Aubrey confesses while stuffing his iPal into his pocket. "What's the big guy?" The flight engineer gets a gleam in her eyes. "The orbital airship...she's over a mile long and all mine. I'm the Chief Engineer on the next run to orbit."

Commanders think ships are theirs, but the truth going back to the days of sail is the chief engineers are the real masters of their vessels. Aubrey, trying to keep the "size matters" jokes out of his head replies, "Then I know I'll be in good hands."

"Run while you can!" comes a muffled voice from his pocket. Aubrey can only give a sheepish smile.

By "up top," the flight engineer is referring to the suborbital space station at 140,000 feet. At the station, Aubrey will be transferring to another vehicle for the flight into Earth orbit. Instead of a four-minute rocket flight, going to space has become a four-day process, with layovers and a plane change to boot. It has only been a decade since people used rockets, essentially great flying gas tanks with fire in the back, to reach orbit. It was a faster trip back then, but also more dangerous. The cost of the old way was so high that only a few could go. "Just under a decade ago I would have needed to be an astronaut", Aubrey muses. Now, as more people were needed off planet, the learning curve to space travel was a little shorter. Instead of years of intense training, Aubrey took the mandatory two-day course, titled "Living Beyond the Earth: Safety, Comfort and Productivity." The class boiled down to just three things— relax, ask questions, and don't push that button if you don't know what it does. Going to space had become a poster: "All I need to know about space travel I learned in Kindergarten."

"All that's missing is a bag of stale peanuts," Aubrey reflects. The comparison with air travel was clear. Earlier in the last century, aviators were heroes—those brave few who possessed the skill, stamina, and courage to take to the air. Now, ninety-year-old grandmothers can fly across the Atlantic ocean in an afternoon to visit their grandchildren. The only stamina required is for waiting in the airport line. The only courage needed is for eating the airline food.

Thinking about food reminds Aubrey that he still hasn't filled out his flight menu. He settles into the nearest seat at the gate and fumbles in his backpack for the forms. The first leg of the trip is only going to last a few hours. As expected, his selections are not of food, but of categories—standard, vegetarian, diabetic, gluten-free, or Kosher. "Standard" sounds a bit too much like cafeteria "mystery" meat, so he checks the box next to vegetarian to be on the safe side. When he flips the page, he sees a map of the cabin showing the emergency exits and how to use the seat as a life preserver.

So much for the beginning and the end of the trip...how about the middle? He has a two-day layover at the station before the orbital trip. Well, he will have to ask.

He hands his form to the uniformed guy at the counter. Aubrey just smiles as the guy gives him that "you should have done this a week ago via e-mail" look. "So," Aubrey ventures tentatively, "What's the food situation like on the station?"

The counter guy drops into script mode. "There are three restaurants, a cafeteria, and snack bar." He then lowers his voice conspiratorially. "The Italian is the best, but whatever you do, stay away from the sushi place."

"The food is covered, but what I am going to do for two days hanging in the sky," thinks Aubrey. Almost as an answer, he hears a muffled "Hey space man, you going to let me out of this oh-so stylin' pocket?" "Only if you behave, lint-breath." Aubrey sighs and clips his iPal back onto his belt.

Despite the mundane discussions about food, Aubrey is still nervous. Foregoing his seat, he goes to the window. Snow flurries pelt on the glass. In spite of being a space commuter, Aubrey likes to think of himself as an astronaut with the "right stuff." Do you still have the right stuff, if your biggest worry is the questionable sushi bar? Either way, parked outside the window, that big black blimp is his ride.

JP

Chapter 3
Three Part Architecture

The roar is deafening in your ears, and you shake so hard you're about to involuntarily wet yourself. At least you know you won't fall apart. You can't; you're being pushed down into the seat by a force several times your body weight. The only thing that distracts you is the knowledge that several million pounds of fuel are under you and have just been lit on fire. There is no confusion about where you are. Below you, one of the greatest controlled acts of violence ever created is unleashed. There is only one place you could be—in a rocket heading for orbit. I don't know about you, but this doesn't seem the best way to go about this. We have not progressed very far beyond Wan-Hu, all those centuries ago in China.

There is another way. A way to slowly expend the energy required to reach space...to use the atmosphere instead of fighting it.

A jumbo jet and a cargo ship crossing the Atlantic have two things in common. The first thing is how they expend energy. Both the jet engines of the airplane and the diesel engines of the ship run for long periods of time over the entire journey. Second, the weight of the vehicle is supported by the medium it's traveling in: air for the jet, water for the ship. Now imagine you tried to run a cargo ship like a rocket. At the seaport in Liverpool, you would release all the energy for the trip across the Atlantic at once. The ship would shoot out of a terrific fireball and arch gracefully through the sky. It would sail over the ocean carried by its own momentum. When the ship reached the port in New York, it would have another problem—how to stop. Debates would rage on about using wings, parachutes, airbags, or retro-rockets, but no matter how it would be done, it would be a nail biter.

On the plus side, every engineer on the planet can calculate this method as the most efficient way to get from point A to point B. I wonder if dock workers, upon seeing thousand tons of ship hurtling down at them, would care about efficiency. Of course this is silly; no one would ever ship cargo, let alone people, this way. Yet this is exactly how we travel to space now.

How can the trip to space be spread out like the cargo ship or the jumbo jet? The exact same way they do, by expending the energy slowly and using the medium to support the weight of the vessel. If you want to float on the way to space, you will need an airship. If you want to expend energy slowly, you will need an electrically enhanced propulsion system.

The first reaction is normally, "You can't fly to space, there isn't enough air to support the wings." Actually, there is a lot of air up there. Earth's atmosphere extends nearly to the moon. It just gets very, very thin. When the Space Shuttle astronauts look out of their window in orbit, on the tail of the Space Shuttle they see a glow. This glow is from the tail hitting oxygen molecules at high speed. When the Echo 1 balloon satellite orbited the Earth, the front half was pushed flat by the atmosphere.

Both the American Skylab and the Russian Mir space stations crashed down to Earth from the drag of the atmosphere in orbit. A volume of air so thick that it can bring down a 168,000-pound space station is more than a trivial amount. In the simplest terms—if there is enough atmosphere to bring the big stuff down, there's enough to carry the big stuff up. The amount of air in the upper atmosphere is known. The engineering for determining how large of a wing is required is straightforward and so is the calculation for the size balloon needed.

As bad as it is, the danger is not the major reason you and I don't go to space on vacation. It's the cost. Reaching orbit costs around 10,000 U.S. dollars a pound. That's the cargo rate. If you and I want to go, the only ride is with the Russian space agency. The going rate is 20 million U.S. dollars per person.

The fact that spacecraft are large, very technically complex, and operate in a harsh environment seems to justify these costs. However, there is another transportation system that faces similar hurdles and yet has significantly lowered costs—ocean shipping. Ocean shipping operates in an extremely harsh and unforgiving environment. It requires a highly trained and experienced crew on board for months at a time. Modern cargo vessels are amazing high tech machines that bring together all the latest advances in science and engineering, yet the cost of ocean shipping reaches as low as five cents per ton mile. If these ships could go to space, the cost equates to pennies per pound!

What can account for the chasm of this cost difference? It can't be the energy used. It takes more energy to move a millions tons of cargo across the Atlantic Ocean than the Space Shuttle uses to get to orbit. Time? It's not there either; a cargo ship can take a month to cross the ocean compared with a five minute ride to space.

The reason for the cost difference can be found in history. We have been building and sailing ships for traveling on the water for thousands of years. Countless designs, methods, and materials have been tried. Seal skins, wood, steel, oars, paddle wheels, jet turbines, nuclear reactors, and even kites have been employed to move across the waters. With all this time and all this experience, humans have worked out very efficient ways to move cargo across the ocean. With space travel, we've just started dipping our toes in the pool. All that has been tried so far are rockets. Perhaps what is needed is a new approach.

Imagine a research balloon floating at the top of the atmosphere. This balloon is just sitting there, forty miles up. Balloons are beginning to reach these altitudes. What would happen if you gave it a push? Now you've moved it forward. Say you put a slow burning rocket engine on it so you could keep pushing it. Now you have this balloon scooting around at the edge of space. Ok, not very exciting. But now say you give the balloon an aerodynamic shape so that forward motion provides lift.

Now it climbs.

How far can you take this? How high? How fast?

Airship To Orbit (ATO)

One of the great goals of rocket engineers is called Single Stage to Orbit (SSTO). This means that the rocket doesn't drop pieces of itself while climbing. Every orbital rocket today, from the Space Shuttle to the Soyuz drops pieces along the way. They have to. They would be too heavy to reach orbit all together. It's as if along the way, the heavy rocket was changed out for a lighter one. This stepping-stone approach is called "staging." Material science and engine design are not advanced enough yet to do without it. Flying an airship to orbit is not immune to this limitation. Technology is not yet advanced enough to fly an airship from the ground directly to orbit. ATO uses three stages: a first stage airship, a floating airship as a port, and an orbital airship.

Stage One Airship

An airship is not only the final stage of ATO, it's also the first. The stage one airship's job is to go from the ground to 140,000 feet. This is the thickest part of the atmosphere. It will take three to four hours to climb through it.

Figure 3-1: Prototype Stage One Airship

All of the ATO vehicles are big. The first stage airship will be over 800 feet long and 150 feet tall. Its shape when viewed from above is that of a "V." This big airship will carry both people and cargo. Even though it's much larger than historic giant airships like the Macron or the Graf Zeppelin, its weightlifting abilities will be much smaller.

Shortly after takeoff, the airship points its nose skyward into a steep climb. It uses the climb to trade buoyancy for providing horizontal motion. It may seem odd that lift would create forward velocity. The direct example of that is a glider that uses gravity to descend into the air column and generate forward motion. The "V" shape acts as a wing for this upside-down glider. Gravity provides the force and the interaction between the wings, and the air converts the downward motion to a horizontal one. A lighter-than-air can reverse the process. Buoyancy provides the force, and the wings convert the upward motion to a horizontal one. Acceleration from the buoyancy climb can add up to 200 mph to the total velocity. The airship dives upward. This process is called dynamic climbing. This may seem like a wild ride for passengers. On the contrary, it will be no more troubling then leaning back in a recliner, unless of course you look out of the window.

At lower altitudes, the propellers are used mainly for maneuvering. Above 110,000 feet, the airship begins to level off. The propellers take over providing the propulsion. At 140,000 feet, the airship will rendezvous with the Dark Sky Station. The propellers are used for the final approach and for docking.

The Dark Sky Station

The Dark Sky Station (DSS) is a floating transfer facility. The DSS gets its name from the view. The sky is completely black at 140,000 feet. The airship needed to reach orbit is very different from the one needed to reach 140,000 feet from the ground. The DSS makes the transition possible. It is a docking port for both the first stage airship and the orbital airship.

Figure 3-2: Small Dark Sky Station Prototype

This spaceport will be a floating structure over four kilometers in diameter. JP Aerospace has conducted experiments with many floating platform designs. The best configuration found features five buoyant arms radiating out from a central hub. It will house a crew of 20 to 30 people. Five large, cylindrical arms hold the hydrogen, providing the lift. The lifting gas is held in an array of inner lifting balloons. An outer inflated shell supports the structure. The arms are joined at a central hub. The DSS looks like a giant starfish. Accommodations for personnel will be inside, along the base of the arms. The first stations to fly will have a crew of two. The large stations will accommodate as many as 200 people.

Instead of remaining fixed over one location, these facilities will slowly circle the world. They will have sufficient maneuverability so that with advanced planning, they can avoid over-flying any unfriendly country. The DSS never lands. It is a permanent outpost at the top of the sky.

The carrying ability of the first stage airship and the orbital airship are not the same. In a three-stage rocket, all of the stages are optimized for lifting the final payload. The bottom stage does not carry the optimal cargo weight for getting through the lower atmosphere. The final rocket stage is also not optimized rocket technology for orbital insertion. A conventional rocket is a compromise between all the stages. This is a huge limiting constraint. ATO removes that constraint. By placing a transfer station on top of the thickest part of the atmosphere, cargo can be queued, staggered, grouped, or separated. This allows the runs to orbit to be optimized. It's similar to the Federal Express hub system, just a little higher.

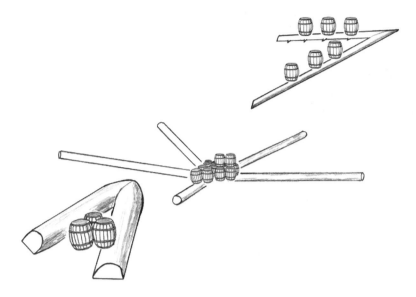

Figure 3-3: Cargo Balancing

Dark Sky Stations will be floating shipyards. Orbital airships will be built and maintained there. Even broken into segments, the orbital airship would not survive the forces of a ground launch and flight through the lower atmosphere. Large, pre-made sections will be folded and compressed, and then carried up by the first stage airship. Once at the station, the sections will be inflated and assembled.

The DSS is a port, a warehouse, and a shipyard. It will be the industrial seacoast of space.

Orbital Airship

The orbital airship starts its journey from 140,000 feet, parked at the Dark Sky Station. Before it has even left the gate, it is at an altitude that conventional rockets would have spent tons of fuel to reach.

It is truly a creature of the sky. After being built in the stratosphere, it will fly from the edge of space to orbit, and back to the top of the atmosphere, never touching the ground.

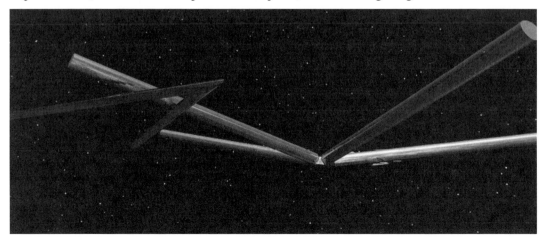

Figure 3-4: Orbital Airship docked at the Dark Sky Station

The orbital airship is the world's largest rockoon. Unlike a conventional balloon-launched rocket, it is both the balloon and the rocket.

We want to move beyond Wan-Hu riding the explosion. To do this, low thrust electrically-enhanced rocket engines will be employed. The great thing about these engines is their amazingly high efficiencies. The bad thing is their low thrust. The thrust is so low that a conventional rocket equipped with those engines would fall out of the sky.

This is where the airship comes in. The airship supports the weight of the rocket. The main use of the rocket engines on the orbital airship is not for climbing. They are for acceleration. But this will not be the acceleration of a dragster. Passengers on board will not even feel it. The airship does use some of the energy to climb.

Some of the lift comes from the remaining buoyancy and the rest is from the large wings. The airship will achieve orbit by performing a balancing act between acceleration and climbing to reach orbital altitude and speed simultaneously. In orbit, the airship can dock with a space station or, if its cargo is a satellite, release it directly into space.

It's now time for the airship-turned spaceship to go home. Small thrusters are fired to pitch the nose up to a steep angle. This causes the drag from the atmosphere on the airship to increase dramatically. The airship slows down and begins to drop out of orbit. By slowing down high in the atmosphere and by having a very low vehicle density reentry, heating is kept very low. The airship uses its engines to return to the Dark Sky Station.

Common Architecture

Building ATO is a major undertaking. It will take decades to fully develop and hundreds of millions of dollars. However, I'm selfish. I want to see it accomplished in my lifetime. Medieval architects would design grand cathedrals that would take generations to complete. No way—I don't want to wait. I don't have that "it will be done after I'm dead" kind of patience.

However, you can't just hope that things will proceed quickly. You must build a rapid method of development right into the structure of the project itself. For ATO, that method is Common Architecture.

Common Architecture is a way to keep the cost and the development time down by sharing systems, components, and design concepts across the ATO system. Development time can be drastically reduced by using common architecture principles. The best example is the inflated struts. Like the steel girders in a high-rise building, the inflated strut is a basic building component of ATO. It is used on the stage one airship, the DSS, and the orbital airship. Even though the struts on each vehicle will be different, the engineers will be familiar with inflated strut technology. They will have the experience of developing and testing many examples. The software for modeling structures is kept to a minimum, as well as the testing equipment, validation procedures, and basic materials to keep on hand. It also focuses the intellectual capital. Everyone on the team, from engineer to manager, will *understand* inflated struts. Wherever possible, the exact same inflated strut will be used across the system.

Common Architecture is similar to Southwest Airlines flying only one type of airplane, the Boeing 737. All the pilots learn to fly the same plane, all the mechanics repair the same plane, the inspectors, managers, financiers, executives, board members all understand this one airplane. This one decision to use the same plane for all its flights has simplified and focused the company. It's no surprise Southwest is one of the few airlines making money and even boasts a perfect aircraft safety record.

The cockpit of the first stage airship will be designed to match the cockpit of the orbital airship. To eliminate pilot error from confusion when changing vehicles, some design precautions will be made. Have you ever switched from a stick to automatic transmission car only to keep trying to push in the clutch that isn't there? To keep this from happening on the way to space, there will be empty space in the cockpit. Wherever there is a control or instrument in one vehicle that is not in the other, that space will be left empty. The pilot may still grab for it, but no harm will be done.

These principles are applied even in the early development stages. The first crew module being built for the Dark Sky Station will be reused on the first piloted airship.

Some parts will be identical in all three vehicles. One of the secrets to creating an entirely new technology is to keep the new stuff to a minimum. The telemetry, life support, and computer systems will all be interchangeable. Some parts will be from the same design family but will be tailored to a specific use. Other parts will be different, but will use the same structural materials such as carbon fiber, polyethylene, and UA protected nylons.

If you were standing inside a lifting structure of an ATO vehicle, you could be confused for a moment about whether you were in a first stage airship, a Dark Sky Station arm, or the wing of an orbital airship. The scale would give it away, but everything would be familiar.

Common Architecture order of priorities:

Identical and swappable

Expanded or modified version

Shared base components

Common design type

Shared base materials

Shared concept

Common Architecture is not just for the final system. It is already being used extensively in the development process. Carbon trusses for the Ascender 90 airship are identical to those in Dark Sky Station III. The telemetry system on the rockets used for testing ATO parts is not only an identical design but will be the same one pulled out of the airship the week before.

JP

Chapter 4
Suborbital Space, the Forgotten Sea

Rockets rush through it; we see through it when we look at the stars. At first glance, near space appears to be just a vast emptiness. This is a deception. Near space is a rich and complex realm. Beginning at twelve miles above sea level, it reaches up to 62 miles. A less specific, but more understandable definition could be the altitude above where the highest airplanes fly, and below where satellites orbit the Earth. It's a spacescape of mile-tall plasma volcanoes and ice crystal clouds. It's where meteors burn and scatter material from throughout the solar system. It's not empty, and there's evidence that it isn't sterile either. High above terra firma, the possibility of life exists.

Near space is also the destination of the burgeoning suborbital space tourism industry. Market studies suggest that thousands of people are ready to put down their cash for a rocket ride. Many trusting souls have already bought tickets even though the ships are not yet flying. For purely commercial reasons, it will be important to get the most out of the trip. Just like finding out more about the pyramids before visiting Egypt, learning about near space will enhance the traveler's experience. For safety concerns, it will be important for the operators of these flight to have a full understanding of the realm their passengers will be zooming through. Near space will be home of the first settlements off the surface of the Earth, a new home for humanity. Today, most methods of reaching space fight against the environment. ATO works in concert with it. To get the system to work, we must understand this environment. This strange place is a resource. A resource to explore, utilize preserve and experience.

The following is just the briefest description of this world surrounding our world. It's like trying to describe all the oceans of the world and everything they contain in just a few pages. These are the things I'll be pointing my camera at on the way to space.

The Road

If you want to lie awake worrying at night, think about this. There are no walls between us and the dangers of vacuum and the radiation of space...no solid barrier. The only thing protecting us from the ravages of space is layer upon layer of gases. These invisible blankets around the Earth extend outward for thousands of miles. The Earth's atmosphere extends ever so faintly beyond the moon.

To help make sense of this region, scientists have broken it up into layers. There are two main ways to categorize the atmospheric layers, by either their electrical properties or by their temperature. Sometimes descriptions of the atmosphere use layers from both systems. This is a big cause of confusion. The five primary layers as defined by temperature are the most familiar and the most useful for our needs. Throughout this book, we will stick to the temperature model to define the layers. Of course, the temperature of each layer can vary greatly on any given day. The temperature at any one moment in an atmospheric layer could be completely wrong for the definition. To simplify the process, scientists have compiled tables of information taken from averaged data gathered over many years. This block of data is called the standard atmosphere. This represents a model of the sky. It allows scientists, researchers, and weather forecasters to all keep in sync. Every decade or so the model is recompiled. This allows for changes in the atmosphere and new, more accurate data gathering tools to be reflected in the model. The 1976 model is the one that is used for this book.

Although good for definition, the temperature difference does not even begin to tell the story of these layers. Each is its own unique environment, like the rain forests or the desert. Each has its own wild set of phenomena. Some of the layers are well known. Others are still a mystery.

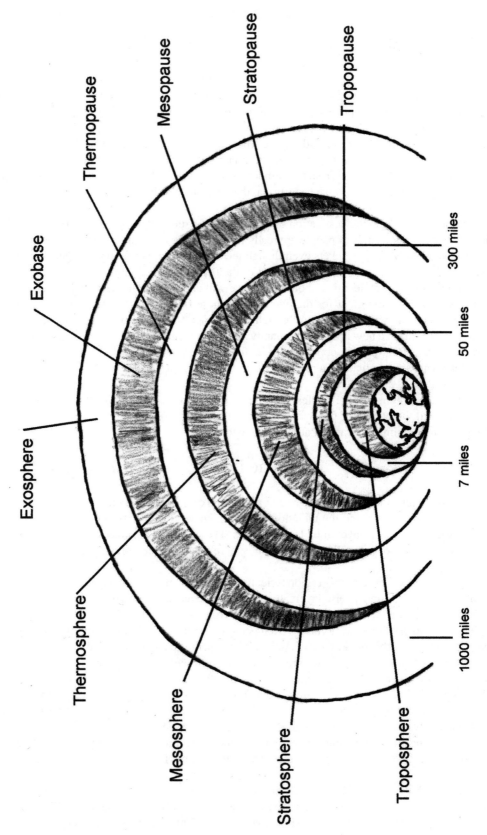

Figure 4-1: The world above our world

Troposphere, where we live

Where do humans live? Many folks would say the Earth. The Earth is a really big place. Human civilization only exists in the paper-thin veneer of thick atmosphere just above the surface of the Earth. We live in the Troposphere. This is where the story starts. For most people, the Troposphere is the only part of the atmosphere they will have any interaction with. When we take in a deep breath, we are inhaling a piece of the troposphere. Clouds, rain, hurricanes, and warm sunny days are all events of the troposphere. It is the first atmospheric layer and the one we are most familiar with.

The Troposphere starts at the ground and extends to approximately nine miles high at the equator. At the poles it drops down to approximately five miles high. In the Troposphere, the temperature constantly drops as you get higher. There's no getting out of the Troposphere on foot. Even at the top of Mount Everest, you are still in the Troposphere.

This layer is a rough-and-tumble neighborhood for ATO. A large part of the system is used just to get above it. When flying a mission, the knot in my stomach always eases when we climb above 48,000 feet and leave the Troposphere behind.

Tropopause

The Tropopause is a buffer region. It is a narrow separator between the Tropopause and the layer above it. The combination of the Troposphere and the Tropopause make up the "lower atmosphere."

Stratosphere

If there ever was a celebrity atmospheric layer, the Stratosphere would be it. Cars, planes, and trains have borne names like Stratoliner, Stratocruiser, Stratojet to imply speed, grace, and high tech. Balloonists in the early twentieth century "defied death in the Stratosphere." The Stratosphere even has a hotel named after it in Las Vegas.

The Stratosphere picks up where the Tropopause leaves off, and reaches to 31 miles high. The temperature of the Stratosphere actually increases with height. At the top of the Stratosphere, the temperature warms up to a toasty 26 degrees Fahrenheit.

Here is the home of the ozone layer. Ozone is a critical protector of life on our planet. Yet, it is a very simple thing. Ozone is an oxygen molecule with three atoms instead of the usual two. These fat molecules block dangerous ultraviolet radiation. The ozone layer is not a separate layer of the atmosphere, but a part of the Stratosphere. Around the equator the ozone layer drops very low, almost into the Tropopause. The ozone concentrations climb to the top of the Stratosphere near the poles.

Stratopause

Here's another one of those funky atmospheric "pauses." These in-between layers tend to be ignored. Even more than the rest of the ignorosphere! These transition layers are even left out of many atmospheric references. At the "pauses," the cycle rising or diminishing the temperature is halted. These stable regions of temperature exist between each of the primary layers. We skip these layers at our scientific peril. The most familiar in-between area on Earth is the shore, the zone between the land and sea. The environment at the shore is very different from both inland and out to sea. Transition zones can be very interesting places. What waits from us in these "pauses?" We have little idea. They have yet to be fully explored.

Mesosphere

Some people have a favorite color; I have a favorite atmospheric layer, the Mesosphere. It is the first layer where "no man has gone before." Although astronauts and cosmonauts have flown through the Mesosphere hundreds of times, they haven't really been there. It would be like saying you've been to St. Louis because you passed it on the highway. The Mesosphere begins at 31 miles, just within range of the highest flying balloons. It extends to 53 miles. The Mesosphere is defined by a turnaround in temperature, which begins dropping. It is the coldest layer of the atmosphere with temperatures reaching down to negative 135 degrees Fahrenheit.

"Outer space" used to start in the Mesosphere, right at the 50-mile mark. Nothing actually happens at the line. It was just a nice happy round number. With the conversion to the metric system, the border to space went up! The beginning of space was moved to 100 kilometers, or 62 miles, another nice round happy number. At the time, I was putting the finishing touches on a rocket in an attempt to fly the first amateur rocket to space. The test flights were just completed when the goal was moved twelve miles further away!

Not much is understood about the Mesosphere, and it happens to be ripe with interesting phenomenon. Sprites, blue jets, and microbial life are just some of the amazing things that occur here. When you see a shooting star burning across the sky, you are looking at the Mesosphere.

Mesopause

The area stretching from the Stratosphere though the Mesopause is known as the middle atmosphere. This is where the ATO action happens.

Thermosphere

Up next is the Thermosphere. This layer begins above the Mesopause and extends to 373 miles. The Thermosphere is so named because it is the hottest part of the atmosphere. Temperatures can reach 3140 degrees Fahrenheit. The temperature can be deceptive. You could still freeze to death in the Thermosphere. Each molecule of gas has a tremendous amount of energy; however, there are very few of them. Heating not only depends on the energy in the object, but also how many molecules you contact. The few really hot molecules you encounter would not pass along enough heat to warm you up. This is another indication that suborbital space is a weird place.

One of the few events of the upper atmosphere that most people can identify are found in the Thermosphere. These are the titanic magnetic storms called auroras. They are caused by electrons from space slamming into the atmosphere. Auroras are found in the Thermosphere at an altitude of 50 miles. Most Auroras glow green. Occasionally you can see the extremely high auroras that are red.

There are two atmospheric layers where people currently live. Most of us live in the Tropopause, but a select few live in the Thermosphere. Even though the International Space Station is in space, it is still in the Earth's atmosphere. At any given time, at least two people live there. The Thermosphere is the beginning of the upper atmosphere.

Thermopause

The Thermopause is considered the top of the terrestrial atmosphere by many researchers.

Exobase

This is one of the most important boundary lines for the physics of the upper atmosphere. The Exobase is where the molecules are so few they don't collide with each other. Above this line, solar wind and other particles streaming in from space move clean through. This has major effects on atmospheric physics. Near the Exobase is the Winter Helium Bulge. This is a huge pocket where the amount of helium is a lot higher than the surrounding area. It was discovered using the Explorer IX Satellite. Explorer IX was a twelve-foot diameter balloon launched in 1961. The Helium Bulge moves during the year so it is always above the "winter" pole. The bulge builds and diminishes moving North to South and back again. It does show that even the upper atmosphere has seasons. It's important to account for the drop in aerodynamic drag for satellites and airships flying though the bulge. It's too early to tell if this could be a resource to be mined or utilized in some way.

Exosphere

The Exosphere is the catch-all layer for everything above the Exobase. Depending on the latest study, it extends either to 800 miles or to out past the moon. The difficulty arises because there is no clean demarcation line between the extreme upper atmosphere of the Earth and solar winds and interplanetary gases. The result is a merger between the atmosphere of the Earth and the atmosphere of the solar system. The sky shares this trait of fuzzy boundaries with the oceans. There is an official line drawn where one ends and another begins, but the actual physical boundaries are vague. The interaction between the exosphere and the sun's atmosphere is called space weather. Space weather is a serious concern for astronauts on space stations and satellite communications.

What about the Ionosphere, D layer, plasmasphere, and all the rest? There are many ways to divide up the layers of the atmosphere. A great deal of confusion comes from mixing up these layer descriptions. The Ionosphere is part of a description system that uses electromagnetic activity to define the boundaries. The systems described above use changes in temperature to define the boundaries. Temperature has a more substantial impact on upper atmospheric technology so the system of the stratosphere and all of the intervening "pauses" is more applicable.

High altitude balloons have taken pictures from the Mesopause. When the light is just right, all the layers are revealed as bands of color. It's a beautiful sight that cannot be seen from the ground nor in space. In my office hangs an eight-foot photograph of the layers of the atmosphere as seen from 100,000 feet. This breathtaking image was taken from an ATO development vehicle.

New Phenomenon

When researchers dove beneath the ocean for the first time, they discovered a whole stack of new and wild things. In the heyday of underwater research, new discoveries were being made on nearly every dive. Those fortunate explorers were truly "going where no one had gone before."

This is the same exciting time we find ourselves in now with near space. Instead of under the sea, the new creatures and wonders are lurking just above our heads.

The Zoo of Lightning

Most people are aware of just one type of lightning, that big forked flash that threatens golfers and gave life to Frankenstein's monster. Move a little ways into near space, and a whole zoo full of different type of lightning emerges.

Sprites

Sprites, so named because they appear ghostlike and practically nothing is known about them, were first identified in 1989 and occur above the stratosphere, into the mesosphere. They seem to be linked with cloud-to-ground lightning and appear as red optical flashes above thunderstorms. The color extends into the infrared region. As would make sense, they seem to have a greater occurrence in areas where the particle concentration is higher, which could lead to more information about the chemical makeup of the mesosphere. They are very short-lived though, and often not detected visually on the ground since you would have to be able to see above the thunderstorm (i.e. if the storm is on the horizon), but they are sometimes seen from aircraft. The direction of the discharge is not yet known; in fact, very little is known at all except that the behavior appears unlike the type of discharge (lightning) we are used to. Their effect on the atmosphere is of course also not known, and there is a possibility they leave residuals in the atmosphere which scientists are especially keen to study. Sprites appear as both quickly shifting clouds and as flashes in groups. Besides being studied by capturing spectral information, these guys emit some low-frequency waves, which can be picked up as radio waves. High resolution photographs are revealing that sprites have an intricate internal structure. We are just beginning to learn about these magnificent and powerful events.

Blue jets are cousins of sprites. Blue jets launch from the top of thunderstorms and then extend all the way to the bottom of the thermosphere 53 miles high. They may be the tallest things on Earth. They travel at 225,000 mile per hour. Scientists suspect that the upper atmosphere and the Earth make a complete electrical circuit. Blue Jets may be a significant upward connection in that circuit.

Tendrils

Tendrils are long strands of light and electricity that sometimes extend down from sprites. Sprites and tendrils, in combination, are the bad B-movie of near space. They look like an eighteen-mile-tall flashing space jellyfish. Tendrils can move down from the base of a sprite at 31,000 miles per hour.

Beads

On occasion, bright beads appear within the tendrils. Beads are approximately 80 meters in diameter and are brilliant white. Beads live longer than the tendrils they form in. However, even their long life span is under a second.

Trolls: Transient Red Optical Luminous Lineament

Trolls are found in the aftermath of really big sprites. They are red spots that form in tendrils and rapidly move downward.

ELVES

Not only are elves mythical beings but they are also a high altitude electrical event and a very strained acronym. ELVES stands for "Emissions of Light and Very Low Frequency Perturbations from Electromagnetically Pulsed Sources."

This is a fancy way of saying "a big glowing blob that we don't know much about." Elves are the highest of these exotic entities. At 100 kilometers up, they have been observed in photographs from the space shuttle. Elves appear as large flat rings of light that rapidly expand and disappear.

Gnomes

Gnomes are another in this strange family of electrical phenomena. Their massive size, bunching volts in the billions, make them more akin to "thunderbolt of the gods" rather then fanciful miniature woodland people. They appear to be large lightning-like discharges that shoot up from large thunderstorms. The power of gnomes makes conventional lightning look like a spark in comparison. They are physically smaller then sprites. Gnomes appear at much lower altitudes, from 60,000 to 100,000 feet.

TGF

Terrestrial gamma-ray flashes are bursts of gamma rays that shoot up from thunderstorms. They were accidentally discovered by scientists using the Compton Observatory satellite to detect gamma rays from space. Only about 70 such bursts have been observed. It is still a mystery how thunderstorms produce gamma rays. The universe shoots gamma rays from every corner at the Earth. It's only fair that occasionally we shoot back.

Mesospheric Ice Clouds

Both a new and old phenomena to be "discovered" in suborbital space are Mesospheric ice clouds. They have actually been seen for many years. What is recent is the discovery of what they really are. As with many new finds a single name has yet to emerge. These clouds are referred to as Noctilucent (night glowing), Mesospheric ice clouds, polar mesospheric clouds, and space clouds.

These are clouds not very different than everyday clouds. What is unique is where they are— over 50 miles up. These clouds are in space. This is one of the new discoveries that you can see for yourself. Maybe you already have. They can now be seen as far south as Virginia. They can be seen glowing with an electric blue hue.

There is controversy regarding how Mesospheric ice clouds get to space. Are they created there, or are they the result of riding currents upwelling from below? The shape of the clouds is reminiscent of lenticular clouds. These lens-like clouds can form in arcs in mountain waves. This fits with the upwelling theory; however, no one has been able to find a way to detect these upwellings. If they are relatives of the mountain wave, it would create a stir in the sailplane community. Sailplane pilots use the mountain waves to soar up to 50,000 feet. Could they use the Mesospheric ice clouds to find upwellings to soar to space?

Understanding these clouds may have a significant impact on the knowledge of climate. They are viewed by many researchers to be an early indicator of the health of the atmosphere—the canary in the coal mine for the entire Earth. Another startling possibility is that Mesospheric ice clouds are triggered by the Space Shuttle. A strong correlation between Space Shuttle flights and the appearance of these clouds has been shown. The Space Shuttle reentering the atmosphere may act as a high-speed cloud seeder. Whatever the cause, the frequency of Mesospheric ice clouds has been steadily increasing over the past two decades.

Life

It's hard to imagine the clear, seemingly empty region of near space being a home for life. It's even harder to picture near-space teaming with life, yet the current data suggests just that. The upper reaches of the sky may be the largest habitat for life in the whole terrestrial ecology.

In the early 1960's, NASA conducted a search for microorganisms in the upper atmosphere with balloons. Dramatic indications of microorganisms were obtained. Viable cell samples were gathered from 110,000 feet and above. In a sad case of not believing their own data, NASA assumed that their equipment must have been contaminated. The data was set aside, and research was terminated.

On January 21, 2001 at the National Balloon Facility at Hyderabad, India, a balloon was launched carrying a very special sampling device. The device was a cryosampler manifold with stainless steel probes. It was designed to take samples from the stratosphere under extreme sterilized and aseptic conditions. Two duplicate sampling probes were carried. One was analyzed in India at the Center for Cellular and Molecular Biology in Hyderabad, and the other was sent to Cardiff University in the United Kingdom.

Between 65,000 to 135,000 feet, the chambers were opened and the atmosphere was passed through micropore filters. The filters were aseptically removed and samples cut from them. The filter samples were then treated with a cationic dye. Cationic dye penetrates the cell membranes of living cells but not dead cells. Any cells penetrated by the dye will glow when exposed to ultraviolet light. The cells can then be detected with a special microscope. The samples showed evidence of living cells in abundance. A scanning electron microscope was used to closely inspect the spots that glowed. The photographs taken clearly show microorganisms. Each of the filters carried by the sampler was designed to capture biological material of a particular size. The microorganisms found are only five to fifteen micrometers across. Only this one filter has been completely analyzed so far.

How much is up there? With so little sampling, only an informed guess can be made. By calculating the flow rate of the atmosphere over the filters, the amount of microorganism in a given volume can be determined. At 135,000 feet, the data showed seven cell clusters per liter. Given a ten kilometer thick area all around the Earth and a clump mass of 3×10^{-13} grams, the result is 102 million kilograms of biological material. In addition, there is no reason to suspect that the lower, more hospitable regions won't also harbor life as well. When the additional samples are analyzed it is expected that total biomass indicated will be significantly greater.

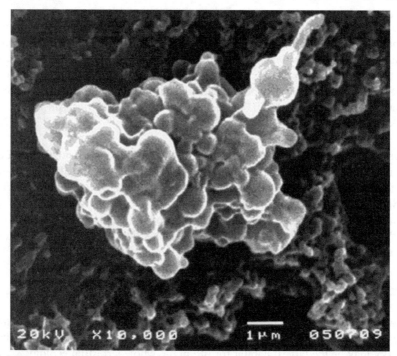

Figure 4-2: Electro micrograph of high altitude sample (Courtesy Cardiff Centre for Astrobiology)

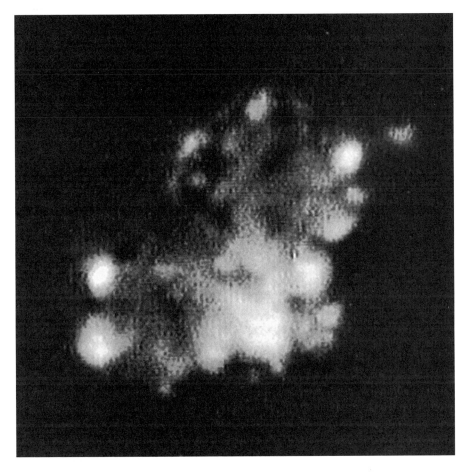

Figure 4-3: Viable Cells (Courtesy Cardiff Centre for Astrobiology)

Life in Clouds

Almost no funds are available for this research, yet it could change our entire understanding of life on Earth. It's like an elephant living next door under a tarp. We know it's there, but no one is taking a peek.

Many studies and a great deal of speculation have suggested that there could be life in the acid drenched clouds of Venus or even in the frozen methane clouds of Jupiter and Saturn. Could there be life in the relatively hospitable clouds of Earth?

Meteor Spectral Findings

There is evidence that life exists even higher up. Scientists in England have evidence that, in part, life is exactly what we are seeing when a meteor burns up.

The Leonid Meteor shower comes around once a year in November. The meteors actually stay still. They exist as a cloud of rocks left over from comets. It's the Earth that comes charging through on its annual journey around the Sun.

Scientists use a spectrograph to analyze the light coming from the meteors. This instrument is used to determine what was in the material that was burning. The spectra of the meteor trail matched that of the heater sample of e.coli bacteria. They concluded that "the meteor trail fingerprint" is most likely due to a population of bacterial particles in the Mesosphere that have been transiently heated to temperatures approaching 400 degrees Kelvin.

Speculation on a Life Cycle

Does life in the upper atmosphere represent a native life cycle or is the life deposited there? The debate goes on that, if not native, does it come from an outside source, such as comets, or does it come from below, riding upwelling winds?

Like the oceans or the rain forests, the sky is a living system. From low altitude clouds to space, knowledge of individual upper atmosphere phenomena is growing. However, there is a real lack of understanding of how all the pieces fit together. The development of ATO will both require a greater understanding of the upper atmosphere and provide the tools for discovery.

The Tools of the Trade

Balloons are still the main tool for exploring near space. Daily weather data is still gathered mainly by balloons. Thousands of balloons are launched every day. They provide the direct measurement that airline pilots use to plan their flights and the information your local weather reporter needs to tell you if it's going to rain on Saturday.

Large balloons are used not only to explore up to the upper atmosphere but also the universe beyond. This field is known as scientific ballooning. These balloons can carry thousands of pounds of experiments. Their missions range from exploring distant stars to monitoring the health of the atmosphere. Over the past six decades, there have been advances in balloon materials, instruments, and batteries; however, the field is in a rut. From 100 feet away, you can't tell the difference between a balloon launched in the 1950's and one launched today. It's still a plastic balloon lifting a box on a string. I attend many balloon conferences. Suggestions of alternatives to the balloon-string-box architecture are always meet with scorn and ridicule. I suspect this book will give them nightmares.

Figure 4-4: A research balloon being readied for launch by JPA

Sounding rocket

This is a great example of a mature technology doing real science on the cheap. Along with balloon programs, they may be the only example of that marriage occurring within NASA or any other government space program.

Small rockets have been used for nearly 70 years to explore the upper atmosphere. These probes measure temperature, density, humidity, electrical and magnetic properties, composition, and even take in samples. Much of what we know about near space comes from sounding rockets.

Aircraft

Military aircraft have been patrolling the lower edge of near space for decades. These "old school" workhorses were employed primarily for intelligence gathering and other Cold War activities. Occasionally, they are pressed into service for atmospheric research. These aircraft represented, and in some ways still represent the cutting edge of heavier-than-air aircraft design.

SR71 Blackbird

Airplane buffs consider the SR-71 Blackbird the greatest plane to have ever flown. Even though this aircraft spent a great deal of time at the edge of space, all of its focus was downward. The Blackbird was the premier spy plane of the Cold War. There have been some interesting anecdotal stories from this aircraft—bugs hitting the windshield twenty miles up for instance. However, even the research version used by NASA was mainly concerned with the technology of flying at high speed rather than studying the environment around it. It is hoped that as time goes by, the Blackbird missions can become unclassified. There is likely to be priceless data about the edge of space wedged between the cracks of Cold War daring-do.

U2, the Dragonlady

Showing that there is more than one way to solve any problem, the U2 spy plane, the Dragonlady, and its successor, the TR3, could not be more different from the SR71. Yet they both fly at the top of the atmosphere, performing the same role. Instead of blasting through the sky at Mach three, the U-2 and TR-3 fly at subsonic speeds, supported by long glider-like wings. Ironically, both the U2 and the Blackbird were designed by the same man, Kelly Johnson of the legendary Lockheed Skunkworks.

NASA maintains a U2 for high altitude research. In a recent study, it made observations of the early solar system by collecting dust samples in the Stratosphere.

MIG-25 FoxBat

The MIG-25 Foxbat was the Soviet Union's answer to the Blackbird. Foxbats were still in use as recently as 2005 in India as reconnaissance aircraft. Among the training equipment used at JP Aerospace are two high altitude pressure helmets used by Foxbat crews as they tried to intercept the Blackbird. Little is publicly know in the west about the Foxbat missions. They are another potential source of great information about operating in this region. Much of the information resides in the minds of those who flew those missions. It is hoped that this data can be captured before it's lost forever.

New Kids on the Block

Scaled Composites

In the Mojave desert in southern California is a company with the unassuming name of Scaled Composites. The company founded by renowned airplane designer Burt Rutan has created some of the world's most amazing flying machines. They built the Voyager, the first plane to fly nonstop around the world without refueling, and Spaceship One, the first private spacecraft to carry someone to space. In the hanger at Scaled Composites, you're likely to find aircraft with wings half swept forward and half swept back, planes with bigger wings on one side than the other and other strange creations. One of their newest aircraft is the Proteus. The Proteus is designed to carry out research in the upper atmosphere, or at least the low end of the upper atmosphere. This aircraft can operate at 60,000 feet, carrying payloads weighting 2000 pounds.

AeroVironment

The Helios was the first of a new breed of ultra-lightweight high altitude airplanes. Its lineage can be traced from the pioneering work of founder Dr. Paul MacCready, the creator of the Gossamer Condor, the prize-winning human-powered airplane. The Helios is flown remotely by pilots on the ground. It is a huge flying wing. It needs to be large and light to support itself near 100,000 feet. It has a wingspan of 274 feet and only weighs 1,322 pounds. The Helios was made of foam, plastic film, and carbon fiber.

The Helios carried no fuel on board. Its fourteen electric motors were power by electricity generated by solar panels that cover the entire upper surface of the wing. The goal of the aircraft was performing high altitude, long duration flights. It was designed for remote sensing, telecommunication, and surveillance missions. The Helios made several successful test flights. In August of 2001, it was flown to an altitude of 96,863 feet off the coast of Hawaii. Unfortunately, the Helios crashed on a later test flight and sank to the bottom of the Pacific.

Forrest M. Mims, III

Sometimes your don't need a satellite or super airplane to reach into the upper atmosphere. One of the most remarkable scientists in the world is Forrest Mims III. Ham radio enthusiasts and electronics buffs know him as the author of a series of handwritten books and guides on circuits and electronics that everyone with a soldering iron keeps close at hand.

In 1989, Forrest created an instrument call TOPS for "Total Ozone Portable Spectrometer." This instrument measures the total thickness of the ozone layer. This ingenious unit is not placed in a satellite or high altitude aircraft. It's held in your hand and pointed at the sky. The entire unit costs less then $300. Using the TOPS, Forrest discovered that ozone data coming from the NASA multi-million dollar Nimbus 7 satellite was wrong. After much denial and dismissal, NASA engineers discovered that the handheld instrument was right—there was a problem with the satellite and the ozone data it was sending.

The roof of our world is an amazing place. There are so many discoveries waiting just within our reach. Whether you have a rocket, an airship, or just a soldering iron, you can join in exploring this unknown ocean.

Chapter 5
Trip to the Dark Sky Station

The big airship cannot taxi like a regular airliner. The passengers have to go to it. All the passengers dash out of the warm terminal and into a waiting bus. As the bus approaches the airship, Aubrey sees it at a different angle than he had from the terminal. He can see that the airship wasn't "blimp-shaped" at all. It is a giant Vee.

As Aubrey walks through the hatch of the enormous vehicle, he stops and stares. The huge cavern he expected to greet him is not there. Instead, he finds a small narrow cabin. He could have been standing in the cabin of a small commuter airliner. A nudge from a fellow passenger behind him breaks him out of thought. Now present is the all-too-familiar bustle of going down the aisle, looking for his seat number.

Finding his window seat, Aubrey stretches out. The legroom is three times that of an airliner. The steward works his way down the aisle, securing the other passengers. When the steward reaches Aubrey, Aubrey already had his seatbelt buckled. "I'm sorry sir, you forgot your shoulder harnesses. First time?" Not wanting to appear like a rookie, Aubrey smiles and says, "No, just forgot," as he pulls on his shoulder harness.

After getting settled, Aubrey notices a cloth pouch on the wall panel in front of him. He smiles; whether going to Baltimore or Mars, flying is all the same. In it, he finds those somehow comforting items found on all airlines: the safety card, an out-of-date magazine, an airsickness bag—he hopes he wouldn't need that——and an upscale catalog.

"How odd," he thinks as he pulled out the catalog. "I'm about to soar to an outpost at the top of the sky, and I'm looking through a catalog to pass the time".

At first glance the items in the catalog are the same as on the plane flight to Alaska. Every now and then, there is something that points to the strangeness of the place he is heading. Along with the rechargeable electric razors and non-wrinkle cloths, there are small isometric exercise machines and anti-space sickness devices, and an e-book, "100 Tips for the Zero-Gee Toilet." The thought of the last item makes him wonder if he should stay at the upper atmospheric station and skip the entire problem, tips or no tips.

He finds a flyer that proclaimed in bold letters "Dark Sky Station Entertainment Calendar" and features the following itinerary:

Monday, Lecture: Biological sampling in the Mesosphere, an Industry Prospective. Concord Pharmaceuticals.

Tuesday, Art Exhibit: Impressionism at the Edge of Space.

Wednesday, Robot Wars Semi Finals.

Thursday, Space Jump 2014 World Championships.

Friday, Supersonic paper airplane drop.

Ongoing: Twice daily walking tours. The movie theater shows Star Wars Episode Seven.

The other passengers are reaching up, adjusting their lights and air vents. They are settling in for the commute. "Well," Aubrey thinks stretching out his legs, "Time for me to settle in too."

The flight attendant's voice comes over the speakers. "Our flight today will take just over two hours. The station uses Greenwich Mean Time. You'll want to adjust your watches seven hours ahead."

Aubrey reads that the station continually circles the Arctic. The winds of the Arctic vortex always keep it in motion. "I guess if you cross all the time zones every couple of weeks you can pick any one you want...might as well pick zero," Aubrey thinks as he adjusts his watch.

The flight attendant continues. "Shortly after takeoff, we will be transitioning your seats for the high angle climb. Please keep your hands on your laps while your seat adjusts, please secure all loose objects, and remember your seat belt must remain on for the duration of the flight."

"Two hours to go 24 miles," grouches the passenger next to Aubrey. "Two hours can get you from London to New York or across town in New York" Aubrey replies, trying to make small talk, "I guess straight up is harder". With the jaded commuter sitting next to Joe Astronaut, the big vehicle slowly lifts off the ground.

Aubrey can feel the ship rising beneath him. Something is missing, but he can't say what. It hits him as the passenger at his shoulder sips his coffee. The shake and rattle of an airliner takeoff is absent. The calm is broken again by the voice over the speakers. "Commencing rotation." The cabin slowly begins to tilt upward. Aubrey feels his seat beginning to move. As the cabin pitches up, Aubrey's seat rotates forward to compensate. He looks out the window and sees the world moving. It appears that he is holding still and the Earth is rotating downward. Twenty seconds later, the seat clicks into place.

The biggest difference and the inconvenience is the aisle. With the seats horizontal, the aisle is a steep shaft running the length of the cabin. "I guess this rules out my in-flight lunch," Aubrey sighs. Just then, a flight attendant's head emerges at his elbow. The attendant keeps raising himself up until he is standing next to Aubrey with a tray of lunch baskets. Aubrey's thoughts go immediately to the scene in *2001: A Space Odyssey*, where a flight steward walks upside-down in the spaceliner.

After taking his lunch box and hot chocolate, Aubrey settles in for the ride. He looks out the window and the shock makes him drop his cup in his lap. He looks again out the window. The ground of the Earth is a line running from below to as high up as he can see. They are flying nearly straight up. Thank goodness the cups are spill proof. They must get that reaction a lot.

Chapter 6
Floating Though History

As exotic as it may appear, ATO has truly old roots. From floating cities to mach balloons, it's all been done before. Its beginning is at the very moment humans took flight and is now on the cutting edge of technology.

Nothing says the "space age" like remote sensing. The satellite photos impact how we grow our crops and how we fight our wars. However, instead of being a product of the space age, remote sensing may be 1,500 years old, as old as the Nazca lines.

In 1977, explorer Jim Woodman and balloonist Julian Knott performed an amazing feat of experimental archeology to support the idea that the ancient Nazcas used balloons to make their famous lines. Ancient vases found in the Nazca desert show what appears to be a balloon with a gondola hanging below it. In one of those notions that are obvious in hindsight, but a leap of imagination beforehand, Jim Woodman proposed that the Nazca people used balloons to direct and observe their oversized artwork. Teaming with scientist, engineer, and balloonist Julian Nott, the two set about recreating and flying a Nazca balloon. The balloon was made using materials only available to the ancient Nazcas. The design was patterned after the vase artwork. On a calm morning, the wood-fired 88-foot tall balloon rose above the Peruvian plains. Pilots Nott and Woodman flew to 600 feet. Turbulence brought the balloon back down. As it landed, the balloonists jumped out of the gondola. Lightened, the balloon rose again, this time to 1,200 feet and flew two miles downrange.

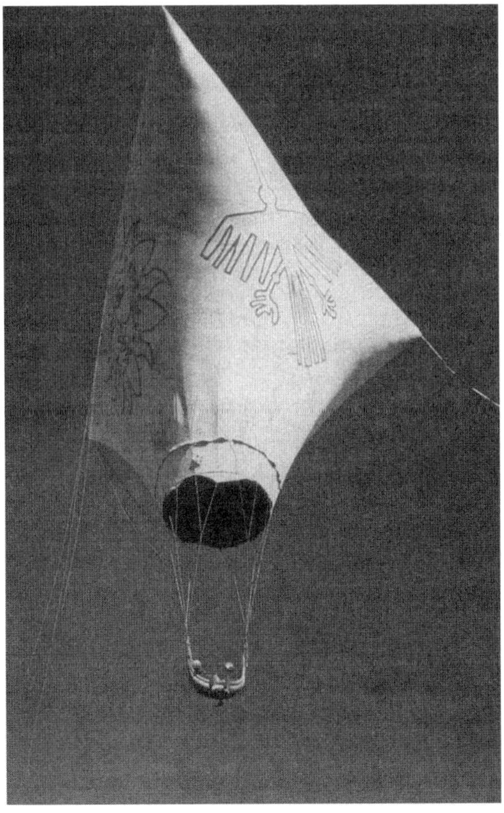

Julian Nott writes (Courtesy Julian Nott www.nott.com):

"When Jim Woodman approached me with his idea that the people who created the Nazca lines could have seen them from hot air balloons I was intrigued but skeptical. Yet we successful flew in a balloon that could have been built by the Nazca people a thousand years ago. And while I do not see any evidence that the Nazca civilization did fly, it is beyond any doubt that they could have. And so could the ancient Egyptians, the Romans, the Vikings, any civilization. With just a loom and fire you can fly! This raises intriguing questions about the development of science and, most of all, the intellectual courage to dare to fly, to dare to invade the territory of the Angels."

It is only speculation as to whether or not the Nazcas flew balloons. Still, it gives me goose bumps to note that near the end of many of the lines, there are burn pits reminiscent of the one used by Nott and Woodman to stoke the fire of their balloon.

Figure 6-1: The Nazca Replica Balloon

Copper and Paper

Roger Bacon truly had the spirit of a Renaissance man; he was just a 100 years too early. A proto-scientist, alchemist, and Franciscan friar, Roger Bacon has even been suspected by some scholars to be the true pen behind William Shakespeare. In his 1250 manuscript "De mirabilis potestate artis et naturae" ("The Wonderful Power of Art and Nature"), he describes both airships and heavier-than-air flying machines. Bacon says about his airship:

> *"Such a machine must be a large hollow globe of copper or other suitable metal, wrought extremely thin in order to have it as light as possible. It must then be filled with ethereal air or liquid fire and launched from some elevated point into the atmosphere, where it will float like a vessel upon the water."*

To the modern ear, "ethereal air or liquid fire" sounds like hydrogen and coal gas. Both have been used to float airships and balloons.

In 1670, Francesco de Lana de Terze designed a flying boat lifted by the vacuum contained in copper spheres. De Lana was the first to conduct detailed experiments to test his concepts. He used this data to create the first design based on scientific principles. To this day, whenever a discussion develops on the Internet about high altitude balloons, vacuum balloons are suggested with a passion. Some ideas never die.

Figure 6-2: Lana de Terze's Airship

Bartolomeu Lourenco de Gusmao has my nomination to become the patron saint of government demonstrations. Bartolomeu de Gusmao was a Brazilian Jesuit priest who studied mathematics in Portugal. While at the University of Coimbra in Portugal, Gusmao built and flew small models of hot air balloons. Documents from the time state the models were of balloons used by the native peoples of South America. This could be further evidence of Nazcas' use of hot air balloons.

In 1709, he demonstrated one of his models to the King of Portugal. The miniature vehicle lifted off the floor supported by hot air from a fire. The balloon then caught on fire. It then proceeded to catch the King's drapes and furniture on fire. Being the seventeenth century instead of today, no one was sued. The King was duly taken in by the event despite losing his decor, (1). As a participant in government demonstrations that have gone wrong, I feel his pain. The Nazcas aside, many historians consider this flight the first lighter-than-air flying machine.

In the early days of the space program, the Russians flew a dog, and the United States flew monkeys. The crew of the first balloon was a sheep, a rooster, and a duck.

The Montgolfier Brothers, Joseph and Jacques-Etienne, were makers of paper in France. They began in late the 1770's experimenting with hot air balloons made out of paper and silk. Their creations grew larger and larger. On June 4, 1783, their latest balloon was large and successful enough to fly in front of the royal family and a large crowd at the palace of Versailles. In addition to the Queen Marie Antoinette, the America scientist Benjamin Franklin was on hand to watch. Franklin even made a donation to help support the flight.

Starting a tradition of combining exploration with advertising that continues today, the brothers joined forces with the wallpaper manufacturer Jean-Bastiste Reveillon. Thus, the balloon carrying the critters before the King was flamboyantly decorated with blue ornate wallpaper.

The most commonly accepted date of humankind's first lift off the ground is October 1783, when Francois Pilatre de Rozier flew in a large Montgolfiere to 80 feet while tethered to the ground. Two months later, de Rozier and Marquis Francois d'Arlandes made a free flight to 500 feet.

South America once again played a big role in floating technology. Alberto Santos-Dumont, the son of a Brazilian plantation owner, began experimenting with airships at the turn of the 20th century. The airship had already taken to the sky; inventor Henri Giffard created a steam-powered airship which flew in 1852. However, it was Santos-Dumont who turned this technological curiosity into a practical flying machine. By adding a gasoline engine and thorough step-by-step development over many airship configurations, he perfected his flying machine.

In 1901, he flew an airship around the Eiffel Tower to win the 100,000-franc Deutsch prize. This dramatic and very public show inspired an entire generation of airship builders and experimenters.

The 1930s saw a rush of activity in the stratosphere. People all over the world were setting off for the high atmosphere. It was the new frontier. These pioneers flew to the edge of space without computers, spacesuits, and having only crude life support systems. They were truly flying into the unknown.

In Germany, on May of 1931, Auguste Piccard and Paul Kipfer make a breakthrough flight. Instead of a basket, they flew in a closed metal sphere. This marked the first use of a pressurized capsule. Before Auguste Piccard, 30,000 feet was a killer barrier. This is where the atmosphere can no longer support human life. The sphere kept in the pressure. They carried an oxygen bottle on board and used alkali to remove the carbon dioxide they exhaled. A 495,000 cubic-foot balloon filled with hydrogen carried them to an altitude of 51,775 feet. They opened the door to space. Auguste Piccard later went on to design the Bathyscape, a metal balloon filled with gasoline with a metal sphere hanging below, for exploring the deepest parts of the ocean. In his life, he traveled to the top and bottom of the world.

Figure 6-3: Auguste Piccard in his gondola   **Figure 6-4: Piccard in flight**

In September 1933, Russian aeronaut, Pilot and Aeronautical engineer, George Prokofiev, Constantin Gudenoff and Ernest Birnbaum boarded their pressurized gondola in an attempt to set the world altitude record. This was their ninth attempt. They took off from Moscow and flew to 62,340 feet. Their perseverance is a lesson for those braving the frontiers of space today. This was higher than anyone had gone before. Sadly, their record was not recognized by the world. At the time, Russia was not a member of the FAI, the record-keeping body.

Undeterred, the team set out for even higher altitudes. Just four months later, on January 30, 1934, the team reached 72,000 feet. On descent, their gondola broke free from the balloon. The three were killed when the gondola struck the Earth.

While Auguste Piccard was flying in Europe, his twin brother Jean was setting altitude records in the United States. On November 20, 1933, Jean Piccard designed a sphere and balloon similar to his brother's called *The Century of Progress*. A Navy officer and a Marine flew on the first flight, setting the official record by climbing to 61,000 feet. The balloon also carried a range of scientific experiments, from fruit flies and a telescope to an infrared camera brought along to study the ozone layer.

The Century of Progress was an early reusable launch vehicle. It was pressed into service again, this time, Jean Piccard and his wife Jeannette, who was a scientist and a teacher, both flew. Jeannette Piccard was the first woman to fly to the stratosphere. She would hold the record for the "highest" woman until rockets carried other women into space. Their flight on October 23, 1934 reached 57,579 feet. The balloon flight carried another aeronaut— Fleur de Lys, a turtle. This record for turtle kind held until turtles on the Soviet Zond spacecraft flew around the moon in 1968.

The legacy of the Piccard family carries on even today. Don Piccard, the son of Jean and Jeannette, was the first licensed hot air balloon pilot in the United States. Jean's grandson commanded the Breitling Orbiter on its historic around-the-world flight. The Extreme Altitude Project (XAP) is the latest Piccard endeavor. Managed by Don Piccard, it is an attempt to fly a person into the mesosphere. They must have helium in their blood.

Project Strato-Jump

On Feb 2, 1966, Nick Piantanida flew to 123,500 feet, setting an altitude record that still stands today This space-suited hero was not a NASA astronaut, not an Air Force test pilot, or even a Navy balloonist. He was a truck driver from New Jersey. He raised the funds, planned the missions, and then flew them. In 1966, he was running his own space program. Piantanida was aiming to sky-dive from the edge of space. He flew three flights all together. The first one on October 22, 1965 had to be aborted at 22,700 feet. On the second flight, he reached 123,500 feet, higher than anyone had ever gone before. However, a frozen connector prevented him from jumping from that altitude. He rode the gondola back down. On the third flight, the ground control team heard the sound of rushing air from Piantanida's helmet microphone at 57,600 feet. A second later, the partial word, "Emergenc–" was heard. The ground team sent the command to cut away the balloon. Less then thirty minutes later, the gondola was found.

Nick suffered from a rapid loss of pressure in his suit. The decompression accident put him into a coma. He died six months later. The second flight was the last to reach those high altitudes. Since then, no one has been back. When ATO flies, we will be humbly following Nick Piantanida.

There have been many amazing balloon flights in this interval, Ocean crossings, around-the-world flights, and long duration rides—however, the great heights remain a no-man's land. Instruments yes—people, no.

Rockoons

In 1956, before NASA, before Yuri Gagarin, even before Sputnik, a professor at the University of Iowa was using the atmosphere as a stepping-stone ladder to space. His name was Dr. James Van Allen. He used a balloon and rocket combination called a rockoon to reach altitudes up to 82 miles. A small rocket was attached to a line hanging down from a weather balloon. Bellows attached to the bottom of the rocket would ignite the motor when the rocket reached over 70,000 feet. The rocket would leave the balloon behind to fly to space. Dozens of launches were done this way.

Another early attempt with rockoons was Project Farside. Farside was an Air Force program with the goal of reaching the far side of the moon with the technology available to them in 1957. The system consisted of a four-stage rocket fired from a balloon at 80,000 feet. The first Farside rockoons were designed to carry a four-pound payload to an altitude of 4,000 miles. Six launches were attempted. Only the final one was successful. On the sixth launch, the first stage launched the rocket through the balloon. The remaining stages pushed the rocket to its maximum height of 2,700 miles, farther into space than the space station.

Unfortunately, the project was cancelled before the second phase of the project, that would have reached the moon, was started. Project Farside showed that rockoons could be used to reach extreme altitudes.

Amateur High Altitude Ballooning

The most frequent visitors to the upper atmosphere are amateur radio enthusiasts. Amateur radio operators, known as "hams," have been on the cutting-edge of space technology from the very beginning. When Sputnik flew in 1957, it was hams who listened in. In 1961, just four years after

Sputnik, hams flew the first amateur satellite. There are currently nearly 100 amateur satellites orbiting the Earth. It's only natural that as interest in the upper atmosphere increases, we find hams already there.

Like the professionals, hams use weather balloons to carry their equipment. A typical ham balloon package would include a Global Positioning System (GPS) receiver, a radio, a camera, several locating beacons, and a parachute. The entire package usually weighs less than ten pounds. Some of the more sophisticated flights carry live transmitted video and can act as a temporary communications satellite.

At dawn, the hams gather in a school yard or football field. A ground "mission control" station is set up, the balloon is filled with helium, and equipment is tested. When all is ready, the balloon is let go and the chase is off. While flying upward, communication experiments are conducted, pictures are taken, and data is gathered. Hundreds of missions are flown each year. More amateur radio operators fly to the edge of space each year than any other research company, university, or government. Once at 100,000 feet, the balloon either pops or is released by a command from the ground. During the 20-mile fall, the chase teams will hone in and recover the package after landing. These weekend expeditions are sometimes accomplished by just a few enthusiasts and sometimes by a large community group. This intense amateur involvement is rivaled only by amateur efforts in astronomy and archeology. The experience and knowledge that is being gained will be invaluable as humans reach out into the upper atmosphere.

What is the limit? Just how high can balloons fly?

A typical research balloon can climb from 100,000 to 140,000 feet. The main goal of most of their missions is long duration. Flying higher tends to reduce that duration. The upshot is that the upper altitude envelope is rarely pushed. There is no technical reason why balloons can't fly a great deal higher.

In Japan, efforts to reach extreme altitudes have met with great success. As in the United States, high altitude balloon research is conducted by a division of the space program. The Space Science Research Division of the Japan Aerospace Exploration Agency (JAXA) has been pushing balloons higher and higher. In 2002, they truly went where no one had gone before with a balloon flight to 173,884 feet. The purpose of the flight was maximizing altitude. Over the next several years, the Japanese intend to fly even higher. Their slogan is "60km Altitude"—196,850 feet.

Both the Japanese space program and private U.S. companies are pushing higher and higher. A balloon will break the 200,000-foot boundary in the next few years. Airships will soon follow.

How Fast?

Even the idea of a hypersonic, inflated aircraft is not new. In 1964, the Inflatable Micro Meteoroid Paraglider (IMP) flew 5,000 MPH at 300,000 feet. The IMP was an inflatable glider just under fourteen feet long. It looked like a blow-up hang glider. It was made from three inflated tubes joined at the nose, with fabric in between the tubes making up the wings. A long inflated tube extended downward balancing the vehicle and holding the instruments.

Its purpose was to measure micrometeor impacts during reentry. Engineers were worried that these sand-grain-sized meteors would punch holes in the heat shields of a space capsule coming back from orbit. The IMP would run the gauntlet and count the resulting impacts.

The wings of the glider were the micrometeor detectors. They were made from alternating layers of polyethylene and aluminum foil. The aluminum layers were given a high electrical charge. The wings became giant capacitors. When a micrometeor hit punctured the wings, the vaporized rock would ground the charged layer for just a moment. The telemetry system would count how many times the electrical grounding occurred, and the engineer would know how many micrometeors hit the glider.

The IMP was launched from the White Sands Missile Range on an Aerobee sounding rocket. At a 96-mile altitude, a shaped explosive charge peeled open the side of the rocket, and the folded glider was dumped into space. The glider then inflated and began to tumble. As it entered the atmosphere, it stabilized and began its glide down.

During the meteor impact experiment, the nose cone of the rocket remained attached to the tube extending down from the glider. Much of the weight was in the nose cone. After the experiment was done, it was to be released. However, the separation didn't happen. The nose cone stayed attached. As the IMP flew, the weight of the nose cone and the instruments inside it caused the glider to fly too fast. The extra speed caused extra heat. Before it reached the ground, one wing boom deflated and glider became unstable.

Figure 6-5: Hypersonic Inflated Glider   **Figure 6-6: Wind tunnel tests**

The whole thing landed in the desert and was recovered. In the words of the principal investigator, Dr. William Kinard, "It flew great!" A 16mm movie camera was carried on-board. It had a split view, half of it showing the scene forward, and the other half showing the horizon off to the side. The recovery helicopter pilot who found the IMP noticed that the camera had broken open. He ran over and stuffed the film canister under his jacket, saving the film. The film confirmed the telemetry data showing that the concept of an inflatable hypersonic aircraft worked!

The concept didn't end with micrometeor experiments. A larger version was developed as a crew rescue vehicle for space stations. The idea was exactly like a life raft on a ship. The vehicle would remain folded and stowed until needed. During an emergency in space, the astronaut would climb into the center tube. The glider would then inflate, re-enter the Earth's atmosphere, and land. The development craft was called the "FIRST," for Fabrication of Inflatable Re-entry Structure for Test. It was designed, but never built.

The IMP showed that inflatable vehicles can fly at hypersonic velocities. The little airship also showed it could survive reentry without a heatshield. This long forgotten test shows the way for Airship to Orbit.

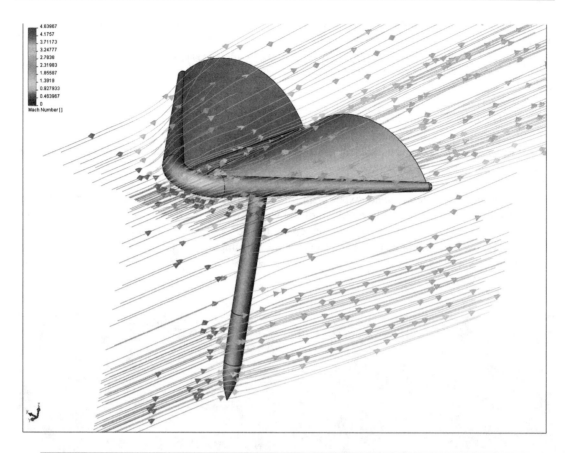

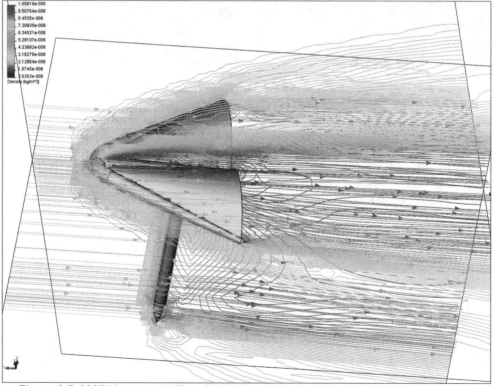

Figure 6-7: 2007 Hypersonic Flow Study Conducted on the 1964 Hypersonic Glider

Beyond Floating: Balloons in Space

From the earliest satellites to an inflated space hotel, balloons have performed missions in space. On February 16, 1961, a Scout rocket hurled a balloon into orbit. Explorer IX was a twelve-foot sphere made of metalized Mylar. It zoomed around the Earth at Mach 24, investigating the density of the atmosphere at 470 miles up. One of its major accomplishments was the discovery of a massive ball of helium sitting over the South Pole just on top of the atmosphere. It is known as the Helium Bulge.

Figure 6-8: Explorer IX

Echo Satellite

When the communication satellites were first conceived, the electronics of the day weren't up to the task. Circuits of the early 1960's were too heavy and drew too much power. Engineer William O' Sullivan, Jr. had the idea of placing a large reflector in orbit and bouncing ground-based signals off it. Echo would be a completely passive system. The large reflector would be a large balloon. The surface of the balloon would be metallized to reflect radio waves.

The idea became a reality with the Echo Satellites. The program was an unprecedented success. The first television image sent through satellite was bounced off the Echo balloon satellite. It enabled a speech by President Eisenhower to be seen live in Europe. Echo One was followed by Echo II and III and the explorer balloons. Miniaturization of electronics made active satellites possible and ended the rein of their passive forerunners.

The Mach Ten Balloon: Project Shotput

As part of the development for the Echo satellite, rockets were used to shoot balloons into the upper atmosphere. This was Project Shotput. The balloons were small, ranging from ten to 20 feet in diameter. They would be ejected from the rocket above the atmosphere and inflated. The Shotput balloon would then reenter the atmosphere at speeds up to Mach ten. When the balloons reached the lower atmosphere, they would burst. The resulting fireworks show was seen by thousands of people along the East Coast of the United States. Project Shotput not only was a precursor to early communication satellites, but it also led to inflatable decoys for nuclear missile warheads. Project Shotput has major implications for ATO. If a fragile mylar balloon can fly 300,000 feet at Mach ten in 1959, surely with today's materials technology, we can do better.

Mercury Balloon Deployment Experiment

When astronauts first flew to space, they tried to deploy balloons. This was the very beginning of science involvement with the manned space program. Among the first scientific research tried in the Mercury program was with a balloon. The balloon experiments were recommended by the Ad Hoc Committee on Scientific Tasks and Training for Man-In-Space, chaired by Dr. Jocelyn Gill.

The balloons were 30 inches in diameter and made of Mylar. The balloon and inflation gas bottles together weighed two pounds. The balloons were to be strung out on a 100-foot nylon tether. The tether had strain gauges to measure the drag of the atmosphere in orbit. The scientists were hoping to measure the difference between the drag at the lowest part of the orbit (perigee) and the highest part (apogee). The segments of the balloon were painted different colors. The experiment was to test the effectiveness of the astronauts' observations of known colors. One section was white in the daylight but would appear to turn blue at night.

The balloon experiments were attempted on Aurora 7 with astronaut Scott Carpenter and on Faith 7 with Gordon Cooper. Unfortunately, the balloon did not deploy correctly on either mission. The opportunity for an early hint at ATO was missed.

Venus

Balloons have even been used to explore other worlds. Balloons were part of the most ambitious unmanned probes ever flown, the Soviet Union's 1986 Vega missions to Venus. The Vega spacecraft sent a lander down to the surface of Venus, deployed a balloon with instruments to fly in the atmosphere, and then continue on to rendezvous with Halley's comet. If this wasn't complex enough, two nearly identical spacecraft were launched only six days apart and operated in parallel.

Remarkably, all aspects of the mission were a success. The balloons sent back information on the Venusian atmosphere, including the existence of extreme vertical bursts of wind. The Vega 2 balloon sailed around the planet for 46.5 hours, floating west for 6,000 miles. In addition, the lander returned information regarding the surface of Venus, and the Halley's encounter gave a first look into the heart of a comet.

These are the only balloons so far to have drifted in the winds of an alien world.

Figure 6-9: Guinea republic postage stamp showing the three aspects of the Vega mission.

Figure 6-10: Vega Venus Balloon

Decoys and targets

Supersonic balloons reentering the atmosphere from space have been with us from the early days of the Cold War. Nuclear missiles carried balloons to deploy along with the warheads. These balloons were designed to mimic the warhead and act as a decoy. These balloons reenter the atmosphere at speeds above 10,000 MPH.

Reentry vehicles

Using balloon reentry is not just for nuclear war anymore. Russia and Germany have partnered to develop an inflatable reentry system. The vehicle, called IRDT for Inflatable Reentry Descent Technology, was launched on a Russian Soyuz rocket. After six hours in orbit, it was inflated. It successfully reentered the atmosphere and landed in February 2000. A second vehicle was flown in July of 2002. However, after the launch, the telemetry system went silent, and no further word from the vehicle was received.

The Hotel Balloon

In 1999, hotel magnate Robert Bigelow formed Bigelow Aerospace. His goal is to build and operate inflated hotels in orbit. In 2006, Genesis 1 rocketed to space. Genesis I is a one-third-size scale version of the space hotel. A Russian Dnepr rocket was used to place it into orbit. Upon reaching orbit, the mini-guesthouse inflated and was a great success. The design of the inflatable module was an extension of the NASA's TransHab program. However, unlike NASA, Bigelow followed though and flew his to space. Bigelow Aerospace is getting to launch bigger versions of Genesis I. I would personally love to fly an inflatable launch vehicle to an inflatable space station.

Apollo Balloon

There are many "almosts" with ballooning in space. Everything from inflatable space shuttles to balloon space stations. One of these "almosts" was a proposal to float the booster stage of the giant Saturn moon rocket down with a hot air balloon. The idea was to save money by reusing the booster. After the booster used up its fuel and separated from the rest of the rocket a 275-foot diameter balloon would inflate out of the top. The balloon would act as a decelerator, slowing the booster down during reentry into the atmosphere. Once reaching the lower atmosphere, burners would heat the air, turning the balloon into a hot air balloon. The booster would then be gently lowered into the ocean. The balloon was to be made of glass fibers, a legacy from the IMP glider program, in order to handle the heat.

Figure 6-11: Apollo's "Almost" balloon.

Real Floating Towns

When hearing about Airship To Orbit for the first time, the concept of a floating town strikes most people as "really out there." As unlikely as it seems, people have been living in floating cities for hundreds of years. These communities are not floating at the edge of space, yet they still have something to tell us.

Uros is a collection of dozens of floating villages. They are located in Lake Titicaca in Peru. These villages are really floating islands made of reeds. As many as ten families live on an individual island. There, they raise cattle, fish, and entertain tourists on their islands. The people of Uros bear the same name and have made their floating homes here for hundreds of years. They were originally used to escape from Inca warriors. These floating sanctuaries have been their homes ever since. In a great tie to ancient balloonists and Airship to Orbit, it was on one of these floating islands that the reed gondola was built for the Nazca balloon experiment.

People from all over the world, from Loktak Lake in India to Lake Kyoga in Uganda, have built and live in floating communities. The Dark Sky Station is not really that different from Uros or Loktak, using woven plastics instead of woven reeds. It just floats a little higher.

If floating mats of reeds are not your style, there is an upscale option—luxury liner cruising condos. Several companies offer extravagant apartment living aboard converted cruise ships. Navy aircraft carriers are often described as "Floating Cities." That description is well deserved. Over 3,000 people live and work onboard. Even though the society of the aircraft carrier is a created one rather than natural, it is still a good model to look at for planning for life 24 miles up. Both have doors that can kill you if they're opened. Both are very technically oriented societies.

What is needed for creating a successful culture on a Dark Sky Station is a fusion—the melding of what the Uros know about family life on reeds, and what the Navy knows about running an aircraft carrier.

Chapter 7
Flight and Approach

The constant sense of climbing was reinforced by a small display set under the window that showed the steadily increasing altitude. As the display showed 60,000 feet, Aubrey noticed that the sky was growing darker. "It can't be that late," he thought as he looked at his watch. "No, it's just after 1:00 PM." Then he realized just how high up he was. The sky was growing darker because he had climbed over most of the atmosphere.

He could see the bright star coming across the window every half-hour. It struck him that the airship was like a racing sailboat, only pointed up. "Too late in the day to be Venus," Aubrey thought. Then it hit him, it wasn't a star or planet, it was the Dark Sky Station. The ship seemed to be tacking back and forth. The airship was making its final climb before approaching the station. By taking advantage of wind blowing in different directions at different altitudes, the airship could travel across to the station using very little energy. "Could you tack all the way to the stars," he wondered? He looked down at the Earth during a tacking turn. The combination of being tipped seventy degrees upward and the world making a rotating arc below made his stomach suddenly queasy. iPal was being helpful:

"Hey Astro-Dude, be sure and put me away before you lose it. My warranty doesn't cover the icky stuff."

"Aren't you supposed to be funny?" Aubrey shot back.

"You were seriously misinformed, Commando Cody," was the tiny reply. Aubrey decided not to look out during the turns.

Sometimes the time between turns was close together. Sometimes they were as long as 30 minutes apart. Every third or forth "tack" Aubrey's destination would come into view. For the first hour it was merely a brighter and brighter dot. The first view after the altimeter showed 70,000 feet, he started to see a shape. The last tack was long. When the gentle turning motion of the big ship started, he strained out the window for a glimpse. Instead of a bright spot, an asterisk seemed to glint and hang in space above them. Ten minutes later, the asterisk became a ceiling fan. The sky had slowly lost its blue on the climb. It was now black as night. The strange feeling of being in a different world was amplified by the shining blue below. The unbelievably bright blue under him was not the sky; it was the Earth. His perceptions were turned upside-down. Two hours into the flight, the ship approached its destination.

Aubrey had studied everything he could find out on the station for the past six months. He had seen photos, layouts, and even video of this very approach, but he was still dazed by the two mile-wide city just hanging there in the black sky in front of him.

The pilot's voice broke in on the magic of the moment. "This is Captain Mason, we are on final approach to the station. Thank you for flying Vertical Air, and welcome to the Dark Sky Station."

After a moment, the captain's voice was replaced by the flight attendant's. "We are about to enter horizontal flight. Please place your hands on your lap during seat rotation." Aubrey could see the cabin around him begin to move again as the big ship leveled out. He felt his seat rotate back. Leveling out was much slower then the swift move to vertical at takeoff. When the seat clicked back into place, Aubrey turned to the window. Out there was a skyscraper laying on its side, just hanging there in space.

Aubrey heard the familiar sound of landing gears extending—no, he reminded himself, not landing gears—docking clamps. There was a slight jolt as the ship docked. Aubrey knew this was a busy time for both the crew on the station and the crew of the airship. With the station and the Ascenders both buoyant, docking and undocking would have no impact on the lift. The transfer of people and cargo did change things. As the weight was shifted between the vehicles the buoyancy of each would need to be adjusted. Passengers were unbuckling and getting to their feet. A quick grab into the luggage compartment, and they were off. Aubrey, along with the rest of the passengers, walked on to the city at the edge of space.

Chapter 8
Floating Cities and Monster in the Sky, Floating in Fiction

Science fiction has forecast flying machines, space travel, and travel under the sea. The earliest sci-fi stories had balloons flying to space. Cities in the sky also made their first stage appearance. Few of these fictional floating metropolises are grounded in science; most are just fanciful. Many are great cities or industrial complexes hanging in the sky with no explanation.

In Jonathan Swift's classic, *Gulliver's Travels*, Gulliver ventures to the most incredible places in the world. The most amazing of all is *Laputa*, the floating city. The inhabitants have devoted themselves to art, music, and study. This began a theme of sky dwellers being a cultural elite that carried on to today. Swift wasn't content to have his city just float there through no apparent reason. Laputa was kept aloft by magnetism.

Everyone has seen a picture of ATO, but doesn't realize it. There is one old print that has appeared in hundreds of books about space travel. This seventeenth century woodcut shows a man flying in the air with bottles attached to him. This early astronaut is Cyrano. He longs to travel to the stars. His ingenious plan for getting to space is simple. Early one morning, he gathers up the dew from the leaves around his village. The dew is carefully placed in vials. Cyrano then straps the vials to his arms and legs. When the sun rises to mid-morning, it warms the ground and the dew rises, both the dew on the leaves and the dew in the vials. Our hero is lifted by the vials and carried skyward into the "above beyond."

Figure 8-1: ATO, Seventeenth Century Style

Cyrano De Bergerac was one of the early pioneers in science fiction. One of his trademarks was to star as the hero in his own stories. Not that this author would ever commit such an act of conceit, (see our hero Aubrey in Chapter 2). The story of Cyrano and his dew is always cited to say, "See how silly we were." Perhaps Cyrano had figured it out right all along from the beginning. Is the tale of Cyrano one of flying dewdrops, or one of solar-powered buoyancy as a means of achieving extreme altitude?

On the cutting edge of balloon technology are solar plastic balloons. NASA is developing these balloons for flights around Mars. These balloons are black to take maximum advantage of heating from the sun. The plan is to use a carrier spacecraft to deploy these balloons on the surface of Mars. At night they rest on the surface, taking soil samples and various measurements with its instrumentation. In the morning, the sun heats the balloon and the balloon rises with its instrument package. The wind blows the balloon across the face of Mars to settle again the next night at its new place to investigate. It would likely be a stretch to point out that both NASA's solar polyethylene balloon and Cyrano's vials both used solar heating to provide lift. It is sufficient to say that instead of being an absurd footnote in fiction, Cyrano De Bergerac may have been on the right track all along (ironically he also was the first to use rockets!)

Stratos in *Star Trek*

Upholding the tradition of the heavens reserved for the gods or at least the god-like, Star Trek's Stratos was a place of art, culture, and beautiful people. It was a floating city for the elite of society to enjoy while the workers were left on the ground. Star Trek's floating city appeared in the episode "The Cloud Minders." Stratos appeared to be the Manhattan skyline set upon a cloud. Criminals were tossed over the side as a convenient form of execution; sky jumpers take note. The surface of the planet below was a wasteland and the clouds were the last holdout for civilization. A cautionary tale for our own planet perhaps?

Cloud City in *Star Wars*

In the second movie in the Star Wars saga *The Empire Strikes Back*, we are introduced to the cloud city Bespin. Bespin gives the impression of being New York City stuck on top of a party balloon. There seems to be a theme in science fiction shows; all cities must look like a Manhattan skyline. These cities tend not be very high. Bespin is shown floating among the clouds. The heroes needed an out-of-the-way hideout, and the Bespin mining colony was the answer. The city was portrayed as an industrial center, a place for gas mining, commerce, and trade. With the exception of Darth Vader, this is how you would want such an outpost to be like. Even the notion of gas mining has a tie to reality. A floating facility in the atmosphere of Saturn or Jupiter could collect helium three, valuable for fusion power. A station in Venus's atmosphere could collect a veritable chemistry set of commodities. The other interesting aspect is that Bespin is used as a port—a way station for travelers to get assistance, a jumping-off point for the rest of the galaxy. It is the Dark Sky Station of the future.

In a stark parallel to reality, the Bespin station's director, Lando Calrissian, comments that the key to their success is the fact that they are overlooked by the bureaucracy. Bureaucracy, red tape, and poor regulation is one of the chief problems in the space industry. In the end it is the bureaucracy in the form of storm troopers and a black-helmeted asthmatic that brings the party at Bespin to an end.

As with Star Trek's Stratos, no mention was made of what keeps Bespin afloat in the first place. For that, we need to travel to the floating city of *Venera* on Venus.

Venera, *Silent Invasion*

In Sarah Zettel's year 2000 novel *The Silent Invasion*, the research station *Venera* offers an accurate picture of a floating city. Venera is named after the Russian space program's probes to Venus. Based in the thick atmosphere of Venus, Venera is a gigantic sphere, a buoyant bubble on which a small city is built. The Venera research station houses the women and men who are exploring Venus. In a twist from what an atmospheric station around the Earth would be used for, the Venusian station is used as a jumping-off point for getting down to the surface to explore. Venera boosts a large enough population that political independence is almost possible. With separation from Earth in mind, they strive for self-sufficiency through a closed-cycle system where all of their water, air, and other consumables are recycled.

A Dark Sky Station would be well-suited to play the roll of the Venera station. With Venus's denser atmosphere, a DSS could be made much smaller and still carry the same weight. While high pressures and temperatures make the surface of Venus hostile, the environment in the clouds is downright inviting.

The story within the story of *Silent Invasion* describes a floating alien society, a culture where the entire population lives in floating cities, never coming into contact with the ground. Living on the ground is completely foreign to them.

In reality, a research station floating in the atmosphere would be the ideal way to explore and study the planet Venus. The former Soviet Union knew this when they flew balloons in the Venusian atmosphere.

Cricket, *Hitchhikers Guide to the Galaxy*

In the third book of the *Hitchhikers Guide to the Galaxy* series by Douglas Adams also talks about people in the sky. The entire industrial infrastructure of the planet Cricket has been moved into the sky. In an echo of Star Trek's Stratos, all the rulers and elite have moved up to the floating cities. The "Masters of Cricket" keep trying to destroy the universe. The occupants of floating cities tend to get a bad rap.

Hesperos and Lucifer in *Venus*

In Ben Bova's 2002 novel *Venus*, a billionaire puts up a ten billion dollar prize for someone to retrieve his son from the surface of Venus. Two entrants compete for the prize. Both contestants use hybrid airship/space ship. The airships, named Hesperos and Lucifer are true ATO vehicles. In the novel they deorbit, fly in the Venusian atmosphere and, at least one, flies back to orbit again. Both competitors were private companies.

Krasnogorskii's Interplanetary Ship

In 1913 in his book "On the Waves of the Ether", B. Krasnogorskii outlined a way to travel to space that nails the fundamentals of ATO, buoyant lift to reach above the atmosphere with a slow acceleration from there. In their attempt to reach Venus, a favorite destination of space balloonists, they build a two part craft. The first component looks remarkable like a Dark Sky Station. When launched it climbs several miles up lifted by four large balloons. When it reaches its peak altitude a smaller ship using a solar sail launches off the platform. The crew with their sail head out to the unknown. If B. Krasnogorskii were alive today I would put him on our team.

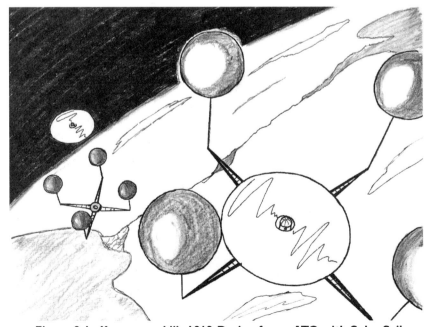

Figure 8-1a Krasnogorskii's 1913 Design for an ATO with Solar Sail

Monsters of the Stratosphere in
Horror of the Heights

It took the author of Sherlock Holmes, Sir Arthur Conan Doyle, to put some fangs into high altitude flight. The upper atmosphere was unknown territory in 1913. It was as mysterious as the surface of Venus seems today. As the early aviators climbed higher and higher, they began to have unexplained accidents. Crashes and deaths were common in the pioneering days of flight, but no one had an answer for these high altitude disasters. These were very fragile crafts, and engine failure, extreme cold, and mysterious ethers were all proposed as explanations. Now we know that the principle danger to pilots at high altitude is hypoxia, the lack of oxygen. Monsters, however make a much better story.

In 1913 Sir Arthur Conan Doyle, wrote the short story that filled the sky with huge vaporous creatures and cloud-like sharks that preyed upon them along with any hapless aviator that strayed too high. The tale told of a diary that was found in the wreckage of an airplane where the pilot was missing. The last frantic entry, written in flight, describes the approaching monster. His descriptions are remarkably similar to those of planetary scientists speculating on life in the clouds of Jupiter.

Figure 8-2: One of Sir Arthur Conan Doyle's Sky Monsters

Even though we have not found giant gas-filled jellyfish with fangs, (I for one, am disappointed), we have found life in the upper reaches. But life at the edge of space is on the scale of bacteria. The impact on our understanding of the size of the Earth's biosphere is immense, even without the fangs.

Mount Flatten, *The Rocky and Bullwinkle Show*

Every fan of Saturday morning cartoons knows the Island of Mount Flatten. Mount Flatten was a floating mountain inherited by Bullwinkle the moose. It was the location of the Upsidaisium mines. Upsidaisium was the mineral that kept the mountain in the air and led to adventures with Boris, Natasha, and mechanical moon mice. It's ironic that the sponsor of the Rocky and Bullwinkle Show was General Mills, the makers of the high altitude balloons for Project ManHigh and Dr. Van Allen's rockoon program.

Sky Boxes in Second Life

Not only in the real world, but floating communities can be found in the virtual world as well. Second Life is a virtual reality community that exists on the Internet. The "residents" log in and live lives in software that mirrors life in the real world. They have relationships, go shopping, and even buy real estate and build houses. Shortly after its inception, the residents invented "sky boxes." Sky boxes are houses that float in the sky. Groups of sky boxes float together, creating villages on the wind. A virtual Dark Sky Station is under construction in this virtual world. It is a unique experiment in science vs. science fiction. Instead of fact following fiction, they are emerging at the same time.

Whether alien or human, about the Earth or on some other distance world, fictional cities in the sky will continue to be one step ahead of the real world. The real world, however, is starting to catch up.

JP

Chapter 9
First Stage Airship

The countdown and "Ignition!" of the rocket launch is about to be replaced by the "Upship" command, followed by a gentle rise. The first step on the road to space is just getting off the ground. Airships have traditionally been low altitude vehicles. A modern advertising blimp has a maximum altitude of only 10,000 feet. The highest-flying production airships were the German dirigibles of World War I. These airships were known to fly up to 24,000 feet to escape being shot down by enemy biplanes. Until recently, no airship had gone higher. The first stage airship for ATO will need to reach 140,000 feet. This is higher than any airship has flown before. This vehicle must be robust enough to withstand the weather of the lower atmosphere—wind, rain, and turbulence—and yet be light enough to reach the necessary altitude.

The key to reaching these heights is in the materials. High strength, lightweight materials are now available. In many cases, these materials have not been used on airships before. Carbon fiber, rip-stop polyethylene, and ultra-thin films are the materials of choice. These materials will be used to create the gossamer structure strong enough to fly to extreme altitudes.

A critical technology needed for the high altitude airship is the propeller. For a propeller to work in the near vacuum at the edge of space, it needs to be extremely efficient. The propeller needs to impart energy from the motor into each and every air molecule encountered. For years the conventional wisdom stated that it was impossible for propellers to function at altitudes above 60,000 feet. New propeller designs have been tested at altitudes up to 98,000 feet. JP Aerospace's own carbon propeller has shown excellent performance when tested to 80,000 feet.

Figure 9-1: High Altitude Propeller (left)

The need for stability at higher velocities and reduced drag during dynamic climbing preclude the use of the traditional blimp shape. An initial design evaluation indicates that a swept wing with an elliptical cross section for the airship's envelope is appropriate. This has been preliminarily validated with the first prototype vehicles. To reach 140,000 feet with sufficient lift capacity, the airship will be approximately 900 feet long. It will have an internal volume of 57 million cubic feet.

The airship will need significant range to rendezvous with the high altitude facility. The problem is fuel and large engines are heavy. To achieve long range and still remain lightweight, the airship utilizes vertical dynamics for forward velocity. To optimize the vertical dynamics, the cross-section of the airship vee arms are airfoils like a wing. An airfoil cross-section is not the best for maximizing volume. The volume of helium you can carry determines the weight the airship can lift. This is another factor driving the airship's enormous size. A conventionally-shaped airship would hold the helium more efficiently; however the velocity and maneuverability would suffer. Electric motors driving lightweight carbon fiber propellers will provide the maneuvering during the dynamic climb phase of the flight as well as main propulsion during the final approach to the floating station.

A common image of an airship is a grainy black and white photograph showing hundreds of people holding onto ropes. It is a strange mix of the Wright brothers and a Macy's Thanksgiving Day parade. In the 1920's, it took hundreds of ground personnel to manage airships. By the 1930's this scene had already changed. When the last of the great airships, the Akron and the Macron flew, ground handling had become a nearly automatic process. The first stage airship would require the same number of ground personnel as a jumbo jet.

When is a dirigible a blimp?

Airships today are one of two types: dirigibles and blimps. The difference is in the structure. A dirigible, like the Graf Zeppelin or the Hindenburg, has an internal rigid framework. A blimp, like the famous Goodyear Blimp, is like a balloon with no rigid structure. The air pressure inside gives the blimp its shape. In the early days of airships there was a third type. This type had the elements of both a dirigible and a blimp. They have been called semi-rigid, DuMont and keeled airship. The term "keeled airship" is the most accurate. Like a blimp, it would "go flat" when it wasn't inflated. However, like a dirigible it had structure—a keel running along the bottom of the gas bag. Both the first stage airship and the orbital airship will be keel airships.

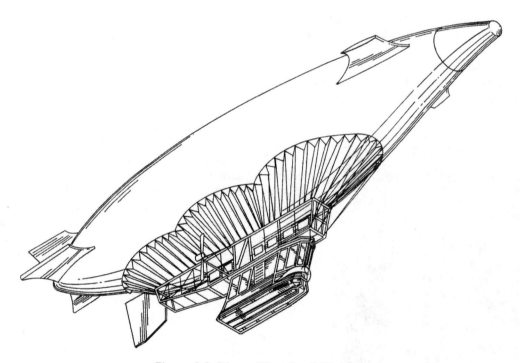

Figure 9-2: Disney "Bad Guy" Airship.

One of the earliest prizes in aviation was won by a keeled airship. Alberto Santos-DuMont was a giant figure in the early days of flight. He pioneered both airships and airplanes. In 1901, he flew his keeled airship "Number Six" around the Eiffel Tower in Paris to win the 100,000 franc Deutsch de la Meurthe prize.

Norge, the famous Italian airship, reached the North Pole in 1926. A keeled airship has even "gone where no man has gone before."

The Norge and airship "Number Six" may have done great things, yet it is another keeled airship that people around the world recognize on sight. Whenever doers of dastardly deeds need to float menacingly above the hero, or a mad scientist needs to flee to his volcanic lab, this airship is there—the Disney Evil Airship. This airship gained its fame appearing in movies such as *The Island on Top of the World*, and *Chitty, Chitty, Bang, Bang*.

The Development Process

ATO development has been a very step-by-step process. The first stage airship development is a good example of this. There have been five prototype vehicles so far. All have been unmanned. Each one has been a step toward space flight.

Ascender 4 (The number indicates the length of the airship in feet)

I have a strong belief that that you don't really get a feel for a design until you build a model and fly it. You can do all the calculations, simulations, and flow studies, but until you glue it together and throw it into the sky with your own hands, you won't understand it. This was the purpose of the little Ascender 4. It floated only six feet above the ground and was powered by rubber bands, but it was the first.

Ascender 16

This Ascender didn't have an outer envelope, just eight four-foot balloons attached to a bamboo v-shaped frame. We used it to explore stability in high winds. The team attached a tether line to the vehicle and reeled it out and up. We immediately knew when we had jumped to the next level. Instead of dancing around like balloons, the little airship was an anchor in the sky.

Ascender 20

Ascender 20 was the first Ascender with inner and outer envelopes. The inner envelopes were made of Mylar and held the helium. The outer envelopes were made of nylon and gave the airship its basic shape. A carbon tube keel ran under both arms, and a carbon crossbeam kept the arms in their proper vee shape. Two small electric motors were mounted on the cross beam for propulsion. This vehicle provided much of the aerodynamic data needed for the larger airships.

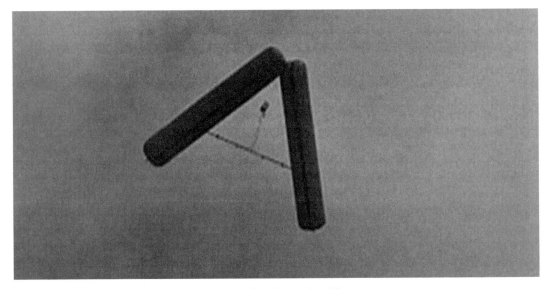

Figure 9-3: Ascender 20

Ascender 90

This airship will always be my favorite. She had all the elements of a full-sized high altitude vee ship. There were ten inner Mylar cells holding the helium. The cells were tied into a gas management system that allowed the helium to be pumped from one cell to another for maneuvering. Two 1/2 hp electric motors provided propulsion. A carbon fiber truss keel held the whole thing together. Ascender 90 was designed to fly to 60,000 feet; however the high flight did not happen. The Ascender 90 was built under an Air Force contract. The Air Force decided to skip the Ascender 90 test flights and move directly to the construction of the next larger vehicle. Ascender 90 became a low altitude testbed.

This poor airship was abused. She was punished for a good purpose. We overfilled some of the lift cells while completely emptying others, causing the structure to twist and contort. We rolled her beyond her design limits to see what would happen. We blew all the emergency vents and watched her crash to the ground. Even though she never flew high, we really shook out the system. We discovered things that have become critical to all of the ATO vehicles.

Figure 9-4: Ascender 90

Ascender 175

If you've never built anything bigger than a 747 before let me tell you, it's an experience. All the lessons of the previous vehicles went into this very sophisticated airship. She was also built under an Air Force contract. The official designation was the NSMV, the Near Space Maneuvering Vehicle. She was intended to be a reconnaissance vehicle looking down from 100,000 feet. The Ascender 175 was destroyed during a high wind accident while being prepared for its first free flight.

Top View

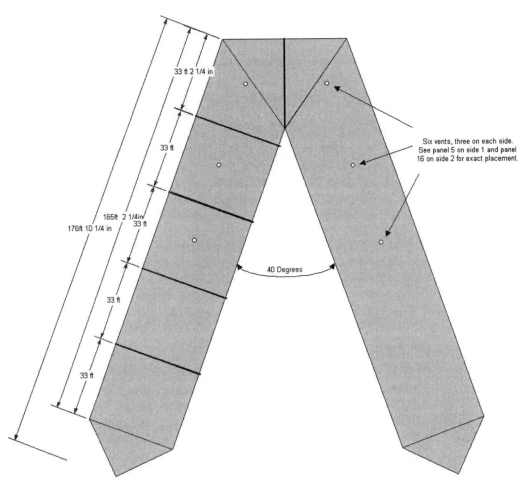

33 ft 2 1/4 in

33 ft

165ft 2 1/4in
176ft 10 1/4 in   33 ft

33 ft

33 ft

40 Degrees

Six vents, three on each side.
See panel 5 on side 1 and panel
16 on side 2 for exact placement.

	JP AEROSPACE			
	NEAR SPACE MANEUVERING VEHICLE V1.0 DEMONSTRATION VERSION			
Vent Locations	SIZE	FSCM NO	DWG NO	REV
		Top View	NSMV v1.0 Vent Placement	5
PROPRIETARY	SCALE	1 : 300	SHEET 1 OF1	

Figure 9-5: Ascender 175

Figure 9-6: Ascender 175 Floating in the Hanger

Figure 9-7: Inside the Ascender 175

What's Next

Ascender 100

The Ascender 100 is a combination airship. It is a test bed for both the first stage airship and the orbital airship. It will be the first Ascender where the arms will have an airfoil cross-section. The leading edge of the arms will be capped in carbon fiber. The first stage airship doesn't need this; however, the orbital airship does. Like the Ascender 90 and 175, it will have a gas management system to pump helium from a cell to any other cell. This gives the airship the ability to rotate around its axis into any orientation. Ascender 100 will have both electrically-driven propellers and electrically-enhanced rocket motors for propulsion.

On all the previous Ascenders, the keel was on the outside under the arms. The keel of the Ascender 100, along with all the other structures, will be inside the envelope. Ascender 100 is under construction. The first test flights are scheduled for the fall of 2008.

Ascender 200

The Ascender 200 is a direct scale up of the Ascender 100 minus the orbital airship components. Ascender 200 will be the chase plane for the early crewed Dark Sky Station missions.

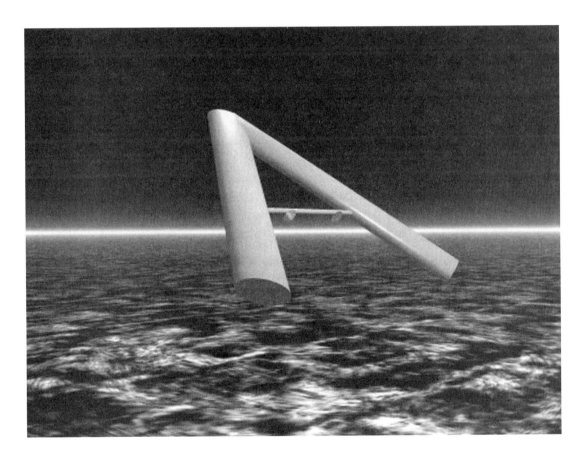

Figure 9-8: Ascender 200

Ascender 900

This is the vehicle that will carry people to the Dark Sky Station. At 900 feet, it will be longer than three 747s parked end-to-end. It will have room for 20 passengers and four crew members. Ascender 900 will be the true first stage airship of the ATO system.

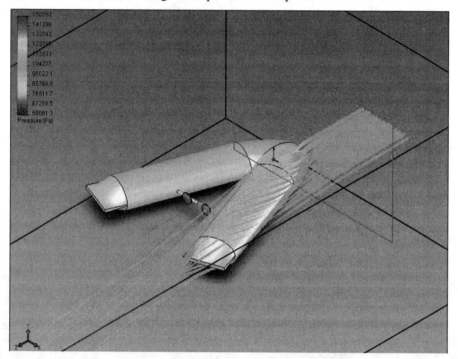

Figure 9-9: Ascender 900 Flow Study

Figure 9-10: The Factory Floor

JP

Chapter 10
Cities at the Edge of Space

Outposts, St. Louis and the Dark Sky Station

When the American pioneers crossed to the west, they did not go all in one step. The technology of the 1800's did not have the ability to carry passengers the full distance from the eastern cities to the west. The Connestoga wagon simply could not carry all the supplies necessary for the crossing. What was a westbound pioneer to do?

They could build a super Connestoga wagon, a huge vehicle big enough to carry all the food, spare parts, medicine, and other supplies to support a non-stop trek to San Francisco. You could almost imagine an eighteen wheel, 60-foot long wagon pulled by a team of 40 horses. Why don't we find these massive prairie ships in the history books? One reason is that they would be expensive. Another is that their weight and higher complexity would make them difficult to maintain and repair. Fortunately, the enterprising pioneers had a better solution; St. Louis. Now, I realize that St. Louis doesn't sound like an invention; however, it was the use of way-points like St. Louis that opened up the American frontier. It is such a simple notion that it is often missed. The ability to reload the wagon, the nineteenth century's equivalent of in-flight refueling, gave the existing technology of the day the reach to travel from Virginia to California.

Current space technology finds itself in the same situation as the early pioneers. You can't reach space in one leap. Unlike the pioneers of the west, today's space pioneers have opted for the super Connestoga method.

The St. Louis of space travel does not yet exist. There's no rest stop or feed store on the way to space. For rockets roaring to space, stopping along the way for supplies or refueling is not an option. For an airship, this is not a problem. A resupply station would work just as well as it did for those covered wagon pioneers. Of course, this outpost is not way out west. It's way up high.

How do you get a city to float at the edge of space? You make it big. Specifically, you give it lots of volume. The visionary engineer Buckminster Fuller designed a flying city made of concrete. Fuller's city was a hollow ball a mile in diameter. The shell of the concrete ball was to be twelve feet thick. Fuller showed that in spite of its massive weight, the ball would float just from the buoyancy gained from the sun heating it. This is very reminiscent of Cyrano De Bergerac's dew vials, just on a larger scale. Fortunately, modern material science has provided many alternatives to concrete. The current estimates for the size of the full scale Dark Sky Station are just over two miles in diameter.

The Dark Sky Station will not only be a stop along the road. Like St. Louis, it will be a destination all its own. Stations will serve as an in situ research station for many fields. Arctic and Antarctic research stations serve as models. In peak research season, hundreds of scientists, researchers, and support personnel live at these facilities. Exploration is conducted in an amazing array of different fields. Experts from all over the world compete for time at these cold polar stations.

High altitude life, astronomy, and anti-matter are just a few of the areas researchers can gain a huge benefit from through the use of the station. Anti-matter studies on balloons have given us incredible new insights. Image what could be gained with a long duration study onboard a DSS. 140,000 feet is an ideal location for astronomical telescopes. It is in effect the world's tallest peak on which to place an observatory. The astronomers would not even need to be present at the station to use the telescope. Only a technical crew would be required. Many telescopes are now operated through the Internet. Scientists in their offices in Princeton as well as on their kitchen tables in Estonia currently share these telescopes. This would be the ideal arrangement for a telescope on the station.

Whenever the concept of a high altitude platform is discussed, without fail someone will ask if they could jump off it. This is not a death wish but a new type of extreme sport, ultra-high altitude skydiving. There have already been practitioners of this ultimate plummet. In 1960, as part of project Excelsior, Captain Joe Kittinger leapt from his balloon-borne capsule at 102,800 feet.

There are several teams attempting to break Kittinger's record. The greatest expense and difficulty for these jumps ironically is not the trip down, but the trip up. A Dark Sky Station would change the nature of these leaps. It changes the nature from a one-time event to a sport. There are approximately 120,000 skydivers in the United States alone. Even jumping from 140,000 feet will eventually grow old. When it does, the orbital airship could provide for even more extreme altitudes to leap from.

The station will also have a place for the arts. Stunning photographs have been taken showing the curve of the Earth shining in the blackness, but no one has been able to set up an easel and paint that scene as it appears to their own eyes. Naturalist artists will often observe and capture elements that the photograph can miss. It's time for the painter and poet to join scientist and engineer in exploring this new sea.

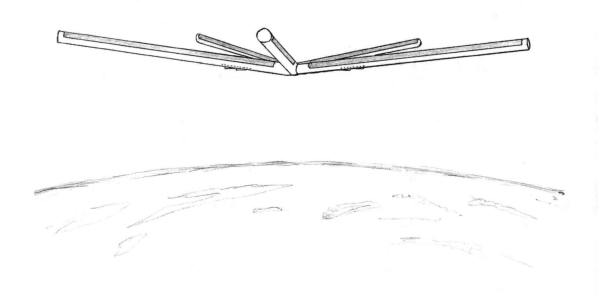

Figure 10-1: Dark Sky Station

Anatomy of a floating City

Many of the parts and pieces of the station are shared with the airships. The basic structural design and lifting systems will look nearly identical. The difference will be the size.

The tremendous size creates equally tremendous issues. How do you build something miles across that flies? How do you make it light enough to float in the near vacuum found at high altitudes? Then, once you've figured out how to build the thing, it needs to be maintained. The tasks of inspection, maintenance, and repair could become completely unmanageable if not for new advancements in materials and in miniature technologies.

A cubic foot of helium can lift one ounce at sea level. At this rate you would need a balloon eighteen feet in diameter to lift a 190-pound person and a balloon 100 feet in diameter to lift a city bus. Helium works the same way a boat does. The lift from the boat comes from the weight of the water it displaces. A balloon displaces air like a boat displaces water. As the balloon goes higher, the air weighs less, so you need to displace more of it. At 100,000 feet, the weight of the air is a 100 times less than it is at sea level. The volume the balloon needs to displace is a 100 times bigger to carry the same weight. The 100-foot bus-sized balloon grows to 470 feet to float at the edge of space.

If volume were the only issue, then a single big spherical balloon would be the ideal shape. However, other factors such as stability, assembly, maintenance, and materials are also critical. The design of the layout of the outpost is a series of engineering tradeoffs. I have been involved with the development of a wide range of floating balloon platforms.

> Criteria: Volume
> Stability
> Maintainability
> Stability
> The ability to assemble at altitude
> Stability
> Segmented lifting volume, and
> Last but not least, stability.

If the platform is not stable, nothing else matters. Stability applies to all phases of flight—launch, climb, and performance at operational altitude.

JP Aerospace has conducted experiments with many floating platform designs. We've flown rings, complex stressed structures, clustered Buckminster Fuller dodecahedrons, and balloons in groups of two, three, six, and nine. We have built platforms that were amazingly stable in flight, but nearly killed us during launch. After years of testing, we had a winner. The best configuration we found incorporates five buoyant arms radiating out from a central hub.

This configuration looks like a giant starfish. The arms from the outside are long cylinders. On the inside, each is made up of a row of Mylar balloons. These balloons hold the lifting gas. The balloons are attached to a keel truss that runs the length of the arm. The balloons and keel are covered by an inflated fabric envelope. As the station changes altitude, the pressure changes the size of the inner Mylar balloons; however the outer shell remains the same. Mounted on top of the outer envelope are thin film solar panels. These panels are the power source for the station.

The Dark Sky Stations are designed for long duration flight. The larger stations are permanently parked in the stratosphere. To bend the old expression, "What goes up, never comes down". The key to long duration is not super new materials, it's maintenance. Many of the materials will weaken over time. Vacuum, ultraviolet radiation, and extreme cold will make major components like balloons last just a few months. The long-term solution is to engineer materials impervious to these hazards. I don't want to wait. The "now" solution is to use existing materials and replace them when they wear out.

The inner lift cells will be the first components to wear out. Even protected by the outer envelope, the inner cells will need to be replaced every hundred days. The station must have the ability to replace the cells in flight. The cells won't be replaced all at once. They will be swapped out one at a time per a maintenance schedule. The lifting gas will be moved from the old cell into the new one during swap out.

Floating Construction Zone

Small high altitude stations can be inflated on the ground and then floated up. Larger complexes will need to be assembled at high altitude. This will be the first step in an entire range of high altitude construction projects.

Building a high altitude station more closely resembles building a bridge rather than an aircraft. Ironically, bridge engineers will tell you that their creations behave more like aircraft than architecture.

A large DSS will not be built from scratch in the air. Large sections will be assembled on the ground. These sections will be inflated with helium and floated. First stage airships will tow the sections to the station construction zone at 140,000 feet. Once there, the segments will be attached to the growing station. The first component to be lifted will be the central hub. The hub will have short arm segments that will provide just enough lift to hold it in place.

Smaller stations will already be floating at the construction zone to facilitate assembly. The initial components will be mated with the small stations to provide stability and some limited mobility. Once the core is in place, additional arm segments will be brought up and attached. As the station gets larger, it will be separated from the smaller "construction shack" station. Assembling the basic structure will take up to sixty days.

Once the structure is complete, the first stage airships will carry up a steady stream of internal equipment as well as liquid hydrogen. The cycle of replacing the lifting cells will begin just as the station is completed. As more advanced materials become available, the time between cell change-outs will grow longer. Once the large station is complete, it will become the "construction shack" for the orbital airship.

The first tests of lifting structural elements have been done. An eighteen-foot truss was repeatedly lifted by balloon. Even though the scale of these tests was small compared to a mile-across station, these tests do show the way.

Helium will be used to lift the station components off the ground. Once at station in the upper atmosphere, hydrogen will replace the helium. Hydrogen needs oxygen to burn. In the upper atmosphere, there is not enough oxygen, so hydrogen will not burn there. Passengers can float at 100,000 feet beneath a mile-wide bag of hydrogen completely safe from fire or explosion. Helium will slowly leak over time through the material of the envelope. Hydrogen will be brought to the station in liquid form to replace these losses.

Location, Location, Location

Being high in the stratosphere is only part of the issue of placement. Above what point on the Earth do you place a floating city? There are three strategies for station placement: 1) station keeping over a fixed point on the globe, 2) allowing the station to drift in a known stable wind current, and 3) a combination of both.

Keeping a station at a fixed point over the Earth requires propulsion—some means of pushing. The two most promising propulsion technologies are ion engines and propellers. A conventional chemical rocket engine uses an explosion to push gas out of its nozzle to provide thrust. An ion engine uses electricity to do the same thing. Ion engines take much less fuel (some can run for a year on a pound of fuel) than conventional rocket engines. The drawback is that they take tremendous amounts of electricity to run. Ion engines that could move a medium-sized station could take gigawatts of power. Fortunately, the surface area of a station covered in solar panels is large enough to provide the power. Ion engines have the advantage that they are solid-state devices, having no moving parts. Also, the technology is well developed. There are currently dozens of satellites flying with ion engines. In addition, one of the most successful space probes in recent years, Deep Space One, is driven by an ion engine. Few ion engines can run at 140,000 feet. The outside air pressure is less than one percent of that at sea level. Many ion engines need much lower air pressure to be efficient or to even run at all.

Figure 10-2: The view from 100,000 feet. This was taken during an ATO development flight

The most efficient propulsion method for the station is propellers. For many years, making a propeller that would work at 100,000 feet was thought to be impossible. I have sat in way too many meetings listening to leading experts extol their proofs that high altitude propellers won't work. To this day it's still claimed to be impossible in spite of several examples to the contrary. There are few greater joys in life than showing a video clip of the impossible happening just after someone in the room has sworn it couldn't be done. In 2003, we developed a propeller specifically for airships operating at the edge of space. The prop was hauled up to extreme altitude by balloon and performed great.

While remaining fixed over a single location may be possible, the tremendous power requirement may make it impractical. Letting the station move with limited station keeping is the best compromise. Like a satellite, a free drifting balloon will orbit around the Earth. An example of this can be seen in flight of the Breitling Orbiter, the first successful around-the-world manned balloon flight. The Breitling Orbiter flight flew around the world primarily near the equator. One of the difficulties the crew faced was getting clearances to overfly all the countries along their flight path. Obtaining clearances from the many diverse "overflight" countries for a floating city or a rocket launch platform would be nearly impossible. Propellers would give the station enough maneuverability to avoid a country's airspace. It will, however, take some advanced planning: "Turn left, Australia is coming up in three days."

Arctic and Antarctic Stratospheric Polar Vortices

Another alternative lies to the North and South. At the poles is a phenomenon known as the arctic (and Antarctic) vortex. This is a continuous wind pattern that blows completely around the world at the northern- and southern-most latitudes. It's like a river that circles the poles. NASA and other research organizations have been using this flow for decades.

The polar vortex is not only an amazing natural wonder, it is extremely useful too. The vortex is a river of wind blowing in a circle around the poles. It is caused by a large low-pressure zone. It is similar in many ways to a large, slow hurricane. It is strongest in the winter, although weaker versions do form in the summer.

A balloon put into the polar vortices will simply circle the continent for as long as it can remain aloft. Countless research balloons that are reported as having gone around the world many times have been taking a short cut, 12,000 miles around the pole instead of 24,901 miles around Earth's equator. This makes an excellent place to park a Dark Sky Station. Parked in the Antarctic vortex, a station would "orbit" the Earth every fourteen days. The path has the added advantage of never being further than 3,000 miles from launch point along the track. The stunning scenery provided to passengers only adds to the appeal.

Horizontal station keeping is only part of the picture. Maintaining vertical position is also vital. At night, the atmosphere cools. This can cause a typical research balloon to descend into dangerous lower winds or even crash into the ground. To maintain altitude, these balloons drop iron filings as ballast to lighten their load. Balloon missions are often limited in duration by how much ballast they can carry. Carrying tons of iron filings would be impractical for a station. The station will overcome the nighttime drop dilemma in two ways: first, by compression of gas envelopes, and second, the station will operate high enough that a limited nighttime descent will have no impact. The station will oscillate between 140,000 feet during the day and 110,000 feet at night.

The Big Pumpkin

Long duration flight has been the Holy Grail of the high altitude balloon community. Literally billions have been spent on trying to make a balloon fly for more than a couple of months. One of the biggest problems is simply ballast. There are only so many sand bags you can carry.

One solution to this problem is not allowing the balloon to expand and contract. This would keep the balloon from climbing and descending as the day and time temperatures cycle. This is accomplished by overpressurizing the balloon. The balloon must be much stronger than a regular balloon to keep from bursting. These are called super pressure balloons. These balloons are much stronger and heavier than conventional balloons.

The largest super pressure balloon project is NASA's Ultra Long Duration Balloon (ULDB), commonly know as the "big pumpkin." The big pumpkin is a massive balloon. It is 450 feet in diameter, and the balloon alone weighs 4,055 pounds. The focus of the balloon is its high-tech skin. Instead of a single layer, the pumpkin is made up of several advanced layers fused together. The first few flights were not promising. On the first flight, the balloon developed a leak at 40,000 feet and had to be brought down. On the second flight, it flew higher but instabilities inside the balloon caused the flight controllers to again terminate the flight shortly after takeoff. The development team kept at it. On January 27, 2005, the ULDB landed in the Antarctic after a record-breaking flight of 42 days. It circled the Antarctic twice, flying in the polar vortex.

Figure 10-3: NASA's Big Pumpkin Balloon (Left - Picture Courtesy of NASA)

Another problem in long duration flight is leakage. Over time, hydrogen will slowly leak out of the lifting cells. If left unchecked, this will cause the station to descend. Again, the Dark Sky Station will not be using an exotic material to solve the problem. Instead, the problem is shifted to one of maintenance. The gas will need to be replaced. Tanks for carrying hydrogen under pressure are heavy. The solution is to carry the gas frozen as a liquid. The frozen liquid gas, called cryogenic, is much more compact as a liquid. This means you need a lot fewer tanks. The tanks, called dewars, for carrying the cryogenics are much lighter than pressure cylinders.

Balloonist Julian Nott was the first person to fly cryogenics on a manned balloon. In 2002, his liquid helium replenishment system flew his balloon over the skies from California to Arizona. He flew his balloon for 24 hours. The gondola carried two dewars of liquid helium and a pressurized cabin.

One of the challenges of using cryogenics with balloons is getting the gas warmed up. Converting it to a gas is easy. However, the gas is still extremely cold. Julian Nott flowed the gas through a series of lightweight heat exchangers, a type of reverse radiator. The flight was a complete success. It showed the real world viability of in-flight lifting-gas replacement.

When I find myself in discussions about advanced balloon technology, the issue of cryogenic lifting-gas replacement always comes up. I wait patiently while my colleagues who "know," rant and rave about how it's impossible. I just smile and wait until they're finished and ask if they had heard of Nott's flight. The odd thing is the response is nearly always, "I don't care if it's been done...it's still impossible!"

Launching

Before you can circle the world, you need to get off the ground. Launch is always a dangerous time for lighter-than-air vehicles. Lightweight high altitude craft are especially vulnerable. Many big balloon projects, including the author's own have been dashed by windy surface conditions. There are several ways to beat the winds-at-launch problem. These solutions pertain to launching the first stage airship as well as the large Dark Sky Station and orbital airship components.

Since the early days of manned high altitude ballooning, sport stadiums have been the place to launch. Not only do they provide a wind block, but stadiums are the perfect place to draw a crowd, sell tickets, and help pay for the flight. In addition to stadiums, open pit mines and large dams have been used for balloon launches. Just about any giant wall, natural or manmade has been pressed into service as a windbreak.

What comes down must go up! Meteor Crater in Arizona is an interesting potential location for launches. At 570 feet it's deep enough to allow setup operations in complete calm. The base is a mile wide. Even the largest components would fit. It is not a prefect location. The lip of the crater is known for high winds. Large station components, like an arm segment would be readied for launch. Then crews would wait for conditions up at the ridge to be right. Meteor Crater sits on private land and launches could be a source of income for the owners.

Bagged Balloon

One common problem with dams, stadiums, and big holes in the ground is they are not very mobile. If wind blocks are not available, or are impractical, one solution is the bagged launch. In a bagged launch, the balloon is placed inside a large fabric bag that is anchored to the ground. At launch, a panel on the top of the bag is torn off, releasing the balloon. A balloon can be launched in very high winds using this method.

The bag also acts as a hanger for the balloon. If there is a delay in the launch, the balloon can remain in the bag until it's needed. If bad weather holds the launch, the balloon can stay in the bag until conditions improve. The bag can withstand winds of up to 40 MPH while keeping the balloon safe. Balloon launch bags have been tested with balloons up to fifteen feet in diameter. They have been used singly and in pairs to lift heavy vehicles. The bag launch technique is very scalable. Even if a large wind block is available, the bag launch will still be the best method for launching large Dark Sky Station components. The balloon bag gives the entire launch process tremendous flexibility. It is a simple, but key technology for ATO.

Figure 10-4: Lifting a truss with a bagged balloon system

Figure 10-5: Bagged Balloon Launch Sequence

Up until now, we've been reviewing passive wind blocks. Tests are being conducted now experimenting with active wind blocking. The concept is to use large, high-speed blowers to create a wind wall. This type of blower is used by fire departments to pull the smoke out of burning buildings. One of these blowers can move over 26,000 cubic feet of air per minute. Arrays of these blowers can create a wind free bubble. Airships, balloons, or stations can be launched from a bubble of calm.

With careful planning and the use of natural and engineered windbreaks, all of the launches required for ATO can be accomplished in all-weather conditions.

Development

Large floating cities and spaceports will be the result of steady, incremental progression: balloon instrument package to platform, platform to temporary crewed lab, lab to long-term outpost, outpost to village, village to city.

This is a version of evolution that is looking at the end result with its first step. Each vehicle along the way is optimized for its task and designed with the final configuration in mind. For example, an unmanned platform may not need a leveling system; however, a future crewed mission will, so the unmanned mission has it. As a result, when the larger crewed vehicle flies, it does so with equipment already flown and tested. There are eight vehicles in the development plan. The first four were unmanned. The next four will have crews.

Dark Sky Station One

DSS One lifted off on May 19, 2001. It had a five-arm, starfish-shaped structure. The arms were trusses made from carbon fiber rods. The trusses were attached to a central hub, also made from carbon fiber. DSS One was 27 feet in diameter. The structure weighed 48 pounds; the total lift-off weight was 60 pounds including the balloons. The station climbed slower than expected. It was flying downrange and starting to reach the limits of the operational range of the ground crews. The station was still climbing when the command was sent to release the balloons. The peak altitude was 46,000 feet. This flight showed the flight worthiness of the DSS design.

Figure 10-6: DSS One On-board Camera

Dark Sky Station Two

The second DSS was a much larger vehicle, 57 feet in diameter. Both stations were designed to use custom-made balloons. These balloons were to be a zero pressure type made from Mylar. A zero pressure balloon is classic research balloon. It is called "zero pressure" because the pressure inside the balloon is the same as the pressure of the atmosphere around it. When it first inflates, it has a small bubble of helium at the top with sheets of loose plastic hanging down. As the balloon climbs, the gas inside expands, filling the balloon. It does not achieve its "ball" shape until it reaches its peak altitude. The custom balloons for the DSS were proving difficult to manufacture. They fell behind schedule. The rest of the DSS One was ready to fly. The decision was made to use latex weather balloons for the mission. Latex weather balloons look like everyday party balloons on steroids.

With the success of the DSS One, progress on the DSS Two leaped forward, and within a few months it was ready for flight. The custom zero pressure balloons still weren't ready for primetime. This was not a problem. The latex balloon worked so well on DSS One, and larger latex balloons were available to lift the heavier DSS Two.

DSS Two was designed to climb much faster than DSS One. DSS One's climb rate was 400 feet per minute. DSS Two would climb at 1,200 feet per minute. This would insure that DSS Two would reach 100,000 feet before being blown too far downrange. DSS One landed 70 miles downrange. We wanted to land DSS Two closer than 30 miles.

The morning of the launch was perfect, not a hint of wind. The crew had been assembling the DSS Two since the night before. At four A.M., the launch checklist was started. DSS Two was a complex machine, and it took just over two hours just to turn it on. At release, DSS Two leaped off the ground. There is an eerie feeling you get watching something that large rushing upward without making a sound. At 1000 feet, something was wrong. The round weather balloons started rocking side to side on the arms. It was suddenly apparent that all ten balloons were rocking in sync with each other, but out of phase, with the balloon closest to the hub swinging to the left while the outer balloon swung to the right. The balloons were rocking so violently that they were actually swinging under the arms. As they wrapped under one of the arms, that side of the station would drop. When the balloons came rushing back up, the arm would wrench upward. After a few moments of this, the sound of snapping carbon rods could be heard from the ground. The station was ripping itself apart. One arm tore completely free. The rest of the vehicle was a tangled mess. It did not however, stop flying. By now what was left of the DSS Two was at 3,000 feet. It continued to drift like a giant shipwreck of the sky. Mission control sent up a command to release the balloons, and with a snap, the nine remaining balloons separated from the remains. Most of the control system was wireless, so even the severed arm released its balloons. DSS Two fell to the ground with a crunch. At impact it made the sound of a soccer team jumping on a thousand cornflake boxes.

Post flight analysis revealed that a low pressure area forms under each balloon. It was caused by, in part, the fast climb rate. The balloons were close enough together that there was coupling between the low pressure zones. We discovered that NASA had seen this problem years earlier and had found a solution for it. Weather balloons used to measure upper winds before a Space Shuttle flight would wobble slightly as they climbed. The wobble would cause the wind reading from the balloon to be slightly off. These errors were too small to matter for normal weather reporting. They were big enough to matter for Space Shuttle flights. A NASA engineer named James Scroggins came up with the idea of molding spikes into the balloon. The spikes break up the airflow around the balloon, preventing the low pressure area below from forming. These balloons are now called "Jimspheres" in honor of their inventor. This experience shows you can never do too much background research. In an engineering twist, we take advantage of the low-pressure zone rather than eliminate it. This phenomenon is used to create stability in two-balloon instrument packages and twin-balloon airships.

Dark Sky Station One showed the inherent stability and excellent flight characteristics of the five-arm configuration. Dark Sky Station Two showed the limitations of the latex balloons for this configuration. The flight provided a tremendous amount of experience in large vehicle construction.

Dark Sky Stations were commercial craft in addition to being research tools. Both carried advertisements and third party experiments.

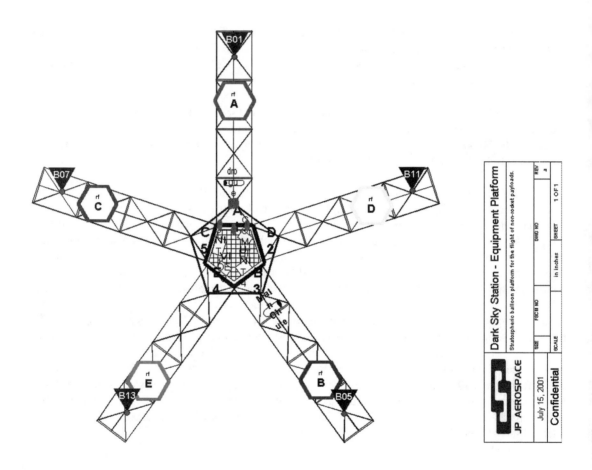

Figure 10-7: Dark Sky Station One Layout

Figure 10-8: Launch of Dark Sky Station Two

Dark Sky Station Three

A third DSS was built in the spring of 2002. It was built for the inaugural flight of the new Oklahoma spaceport. It was the same size as the DSS One. DSS Three incorporated all the lessons learned from the previous stations. DSS Three featured stronger arms, tab assembly connectors, tension cables, and improved electronics.

The morning of the flight, 25 knot winds howled. The DSS flight was scrubbed. Two smaller balloon platforms were flown in its place. The components were so modular that they can be reconfigured to be entirely new vehicles. Dark Sky Station Three was disassembled and became the main structure for the Ascender 90 airship. The central hub is being used for Dark Sky Station Four.

Dark Sky Station Four

The current station under construction is the DSS Four. The DSS Four is a small platform. At 27 feet across, it will be the same size as DSS One and Three. The big upgrade for this DSS is the change in balloon type. DSS Four will fly with zero pressure balloons. This simple change in balloon type has some big engineering challenges. At launch, that tall waterfall of plastic of a zero pressure balloon is beautiful to watch. It is a mess to control.

A research flight flown on April of 2004 used three zero pressure balloons. The mission was called Away 25. It was flown to test a helium pumping system for high altitude airships. As the launch team readied the vehicle for liftoff, the balloons engaged in a swirling dance all their own. The battle to control them continued throughout the launch. It took twelve balloon handlers to untangle and keep the spaghetti of plastic separated. This mess was caused by only three balloons. A DSS uses ten. The options are to sign a contract for the newest professional wrestling league "Plastic Wrestle 3000" or fix the problem. As much fun as it would be to put on a mask and call myself the Ballooninator, the engineering solution was the better option.

The answer to the problem of all the long streaming plastic is to roll it up until needed. The bulk of the balloon is kept on a drum. This way, at launch there is only a small balloon completely filled with helium. The balloon is taut, kept small and low. This makes it easy to manage even in the wind. This concept of a balloon on a drum has been tested. A vehicle called Away 28 (I know, we're not very original in our naming) was built specifically to explore this technology. The initial results were strong. Helium was put in the balloon, and the rig was launched on a tether. The key is in the folding of the balloon. Once the technique was worked out, managing the big wad of plastic became simple and easy.

The roller doesn't just roll the balloon out. It can also roll it back in. If you can roll the balloon back onto the drum, you can replace it in flight. One balloon is rolled in while a new one is rolled out. During the swap, the helium needs to be moved from the old balloon to the new. There are two ways to do this, by pumping or by the toothpaste tube method. Both methods have now been tested in flight. On the Ascender 90 and later on the Ascender 175 airships, the inner balloon that held the helium could be pumped out of any balloon to an adjacent balloon. In the toothpaste tube method, the balloon gets squeezed as it's rolled up. The helium is vented through a duct at the top into the new balloon. This was tested on the Away 7 mission in October 2000. A small balloon was used for launch. As the vehicle climbed, the helium expanded in the small balloon. Soon the balloon was completely full. As the helium continued to expand it was vented through a manifold. The manifold was connected to three additional balloons that were folded and stored in boxes on the side of Away 7. The three extra balloons inflated, and Away 7 continued its climb.

DSS Four will use both of these techniques. Each of the ten balloons will be mounted on rollers. At takeoff, the station will look like a starfish with ten bubbles on its arms. As the DSS climbs, the balloon will be slowly unrolled and allowed to expand to full size. One of the balloons will be connected to a double roller system. Once at 100,000 feet, that balloon will be rolled in, its helium moved, and a new balloon unrolled in its place. Several balloons will be connected to pumps. Helium pumping tests will be conducted throughout the flight.

DSS Four will be a critical milestone for ATO. When city-sized stations are floating at the edge of space, they will trace their roots to DSS Four.

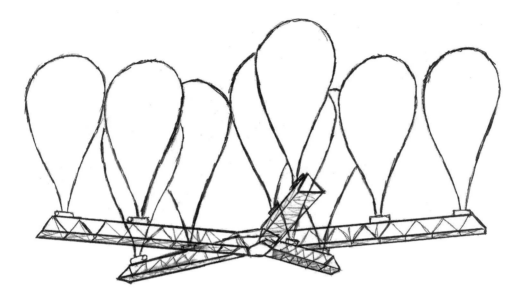

Figure 10-9: Dark Sky Station Four

DSS Block One

Diameter:	200 feet
Crew:	2
Duration:	Four Days
Major Missions:	First Crewed Flight
	High Altitude Rendezvous

DSS Block One will be the first Station with a crew. The crew of two will fly in a seven-foot diameter cylindrical module. For light weight and strength, the crew module will be made out of carbon and boron fiber for strength, and Kevlar for crash protection. The module will be attached to the central hub.

Flights will range from a few hours to four days. The configuration will be similar to that of the smaller test stations, five carbon trusses joined to a central hub. It will use zero pressure balloons and the roller system from DSS Four. Around the balloons will be a fabric outer shell. All the elements of the city-sized station will be present, just on a smaller scale. The Block One station will serve two overall development goals: crewed flight and rendezvous.

The Block 1 station is the small early step. The first flights will be very small. The first mission will be held inside a large hangar, a SSTFF mission (Single Stage To Five Feet). Flights will creep higher, finally reaching 100,000 feet. These missions will more closely resemble the stratospheric flight of the Piccards in the 1930's than Star Trek's Stratos. The missions will reach their peak on the rendezvous mission. On this flight, a first stage airship will fly up to meet the DSS at 100,000 feet. It will be a "bump rendezvous." The airship will approach and make contact, i.e. bump the station. The goal will be to get the two vehicles in the same airspace at the same time.

In addition to flight controls, the crew module will have a life support system. Life support consists of four areas: pressure, air processing, temperature control, and emergency backup. At 100,000 feet, the near vacuum would boil the passengers' blood and burst the cells in their body. To prevent this rather nasty occurrence, the hull of the pod will act as a pressure tank. Air processing involves removing the carbon dioxide and moisture exhaled by the crew, replacing the oxygen used. The carbon dioxide is removed from the air by pumping it though a "scrubber." A scrubber is a set of filters filled with lithium hydroxide. The lithium hydroxide bonds with the carbon dioxide. The carbon dioxide remains trapped in the filter. As the carbon dioxide is removed from the air, the pressure drops. A sensor detects the drop. The sensor is linked to a tank of oxygen. As the pressure drops, the oxygen valve is opened and oxygen is released into the cabin. As long as the tank has oxygen and the lithium hydroxide can absorb the carbon dioxide, the crew will have a breathable atmosphere.

On the way to 100,000 feet, the pod will experience temperatures as low as 120 degrees below zero. You would expect that the main difficulty would be keeping the crew warm; however, just the opposite is true. The real problem is keeping the crew cool. In our everyday experience, objects like clock radios, hot coffee, and ourselves cool through the process of conduction. Conduction is simply the heat leaving one object and moving to other objects that are touching it. When you pick up a coffee cup, the cup heats your hand and now the coffee is cooler. The object that is responsible for most of the cooling in day-to-day objects is the air. This is because the air is touching everything. At high altitude there is very little air. Exceedingly few molecules are there to move heat to. Heat from the equipment and crew builds up but has nowhere to go. Much of the early work on the module was focused on getting rid of the heat

Figure 10-10: Block One DSS with Ascender 200

The crew will not wear pressures suits. We had a great debate whether to suit or not to suit. When the dust settled, the case for not wearing the bulky suits was the clear winner. The margins of safety will be built into the crew module, not the bodies of the crew. At the end of a mission, a small amount of helium will be released and the station will float back to Earth. In an emergency the crew module can be released from the station and descend by parachute.

Upper atmospheric rendezvous is an entirely new operational maneuver. The first rendezvous will be between a crewed DSS and an autonomous first stage airship. The first attempt will be approach only. The two vehicles will simply be in the same air space together. These simple tests will graduate to "balloon bump" maneuvers where the two vehicles will touch envelopes. This lighter than air courting will culminate in the docking of the first stage airship and the station.

The team is currently flying the module, or at least the engineering mockup. Placement of instrument panels, controls, and hatches is being planned out.

DSS Block Two

Diameter:	1,200 feet
Crew:	6
Duration:	Six Months
Major Missions:	First High Altitude Construction
	High Altitude Hard Dock
	Mid-mission Crew Change
	Long Duration Flight
	Transatmospheric Airship Construction
	Around-the-World Flight

The Block Two station will be the first facility constructed in the upper atmosphere. Six segments of the station will be built on the ground. The central hub with short arm sections will be launched first. It will be an empty shell to be outfitted after preliminary assembly is complete. Each of the five arms will be inflated on the ground and floated to 100,000 feet. The components will be lifted from the ground with helium. The helium will be replaced with hydrogen once the station is at operational altitude. High altitude rendezvous testing with the Block One station will be employed in bringing the part together. The crew module will be the size of a motor home. It will feature both an airlock and a docking port. An external deck allows crew members to go outside for construction and to develop the techniques needed for the larger stations to come.

Much of the work of the Block Two station will involve rendezvous. After assembly, the crew will be focused on perfecting high altitude hard docking. This is the next step in upper atmospheric rendezvous. A first stage airship will dock with the Block Two station and transfer crew and supplies. Not only will this extend the mission duration of the station, but also it will be a critical milestone for ATO.

A suborbital version of the orbital airship will be constructed at this facility. This vehicle will be a construction testbed for the giant orbital airship. The Block Two station will become a floating spaceport, launching the suborbital airship and acting as a dock between flights.

In addition to its size, the primary feature of this station is gas envelope replacement and maintenance. The Block Two station will be the permanent station in miniature. All of the activities and tasks required for a permanent presence in the upper atmosphere will be developed and tested.

During their stay, the crew will work on developing procedures for maintenance and repair of every aspect of the station. The crew will don spacesuits and use the airlock to go into the arms of the station. There they will conduct engineering experiments for the larger stations to follow. The station will carry hydrogen to reinflate the lifting cells in case of rupture. The hydrogen will be cooled to a liquid so heavy high pressure tanks won't be needed.

It's possible that tourist operations will begin from the Block Two Station. This station would be the first to carry paying passengers. Paying adventure tourists would not turn a profit for the station, but they would offset part of the cost. By bringing tourists in early in the program, details of how to manage and support them are explored early on. These first intrepid souls would require training and at least partially be part of the crew. Their experience would be closer to those who take "tourist" trips up Mount Everest than those who take an ocean cruise.

Block Two is a transition vehicle, from the experimental to the operational.

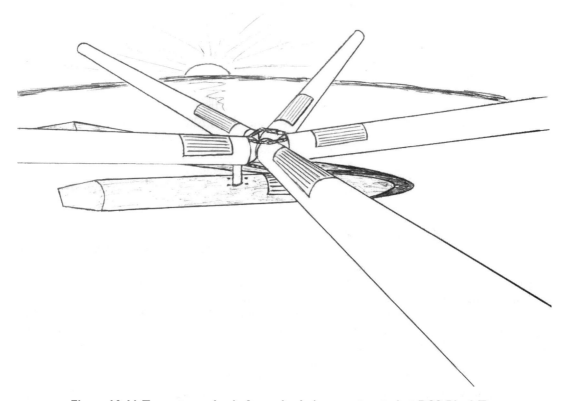

Figure 10-11: Transatmospheric Ascender being constructed at DSS Block Two

DSS Block Three

Diameter:	12,300 feet
Lift volume:	2 billion cubic feet
Duration:	Permanent
Crew:	15
Passengers:	20
Major Missions:	Orbital Airship Assembly
	Orbital Airship Test Flight Support

The Block Three station will be the world's first permanent facility at the edge of space. This huge facility is more like a small outpost than a vehicle. This facility will be constructed similarly to the Block Two station. The modules will be larger, and there will be more arm segments. It will take up to four months to complete this station. The facility has three docking ports. Each port can handle either an orbital or a first stage airship. A first stage airship will be docked at all times for emergency evacuation and for any medical situations that could not be handled on board. The second docking port is for the resupply airship. The last port will accommodate the orbital airship.

The arms will contain fabric tunnels. These tunnels can be pressurized to accommodate maintenance crews servicing the inner lift cells. During cell maintenance, the tunnels will become the world's highest jogging track for crewmembers. The Block Three Dark Sky Station accommodates more passengers than crew. Passengers would include tourists, scientists, artists, the media, and various dignitaries. Passenger revenues will offset operation costs until ATO operations begin.

This is the way station that will make ATO possible. First it will be the construction facility, then the port of call for the orbital airship.

DSS Block Four

Diameter:	18,300 feet
Duration:	Permanent
Crew:	35
Passengers:	70
Major Missions:	ATO
	Tourism
	Station-to-Station High Speed Travel

The Block Four Dark Sky Station is over three miles across and home to over 100 people. The outpost becomes a city in miniature with researchers, maintenance personnel, cooks, artists, and engineers. These facilities will be large enough to accommodate both port functions and temporary guests, like tourists and researchers.

The Block Four vehicle may be at the practical size limit with current technology. To expand, several Block Four stations could fly in formation. Like a small chain of islands, commuter craft would hop between the facilities. In this way, there is no limit to the size of atmospheric populations. These large stations would likely be dedicated facilities—separate stations for construction, science and tourism.

Cities in the Sky - Clusters of Block IV stations

By grouping several Block IV stations together, small cities or villages can be created. Clusters of smaller stations can be an effective way of rapidly building up a large presence in the upper atmosphere without waiting for the development of larger facilities.

JP

Chapter 11
Orbital Airship

At first glance, the orbital airship looks a lot like its little brother, the first stage airship. Indeed, it shares much of the same general design concepts. The keel, inner cell layout, and maneuvering system are the same. The big difference is the details of their aerodynamic design. They operate in very different environments. One moves slowly in thick air, while the other moves fast in thin air.

At the station, the airship is loaded with crew, cargo, and fuel. At release, the pilot slowly moves the airship away from the station. Like a ship at sea, it is critical for the airship to maintain "steerage"—enough forward velocity to provide a minimum of maneuverability. Once clear lift is added to the airship. Liquid hydrogen is flashed into a gas and released into the lift cells. The vehicle, now much lighter, starts to climb. Like the first stage airship, the orbital airship will use dynamic climbing. It will pitch its nose up and climb away from the station at a 50-degree angle.

When the airship reaches 180,000 feet, it is at the maximum buoyancy altitude. This is the height the vehicle could achieve by floating alone. At this altitude, the weight of the airship is the same as the weight of the air it displaces. From here, climbing is accomplished by the wings. As the climb above 180,000 feet begins, the airship has all its mass; however, all of its weight is offset by the lift of the hydrogen. With each foot above maximum buoyancy altitude, the wings bear more of the load. Up until now, the vehicle has been moving subsonic, slower than the speed of sound.

When the airship crosses 200,000 feet, thrust from the engines is increased. The acceleration is so slight that neither the passenger nor the crew will feel it. At 270,000 feet, the benefits of buoyancy are gone. The orbital airship now fully relies upon its wings for lift.

It is not enough flying high. Reaching orbit means reaching orbital velocity, about 17,500 MPH. There are three phases of the orbital airship's acceleration to this velocity: subsonic, supersonic, and hypersonic. Below 200,000 feet, the vehicle has been traveling at a high subsonic speed. The transition to supersonic velocity occurs at just over 200,000 feet. Visions of Chuck Yeager with the "right stuff" exploding through the sound barrier need to be left behind. At 200,000 feet, the air pressure is extremely low. The low pressure greatly diminishes the forces involved. The teeth are pulled from the tiger of the sound barrier.

The next phase is a greater challenge. Supersonic airflow and hypersonic airflow are very different environments. For most of the climb to orbit, the airship will fly hypersonically so the vehicle is optimized for that environment. The slower supersonic portion of the flight will be the most inefficient part of the flight. An example of this apparent contradiction can be found through the experimental hypersonic bomber, the XB-70 Valkyrie. The Valkyrie was driven by six massive jet engines designed to push the bomber to Mach three. Upon the failure of any one engine, the emergency procedures called for running the afterburners on remaining engines. This was in spite of the fact that the speed-boosting afterburners use a tremendous amount of fuel. If the Valkyrie were allowed to slowdown below its hypersonic cruise velocity, it would lose 90 percent of its range. The Valkyrie had to maintain its high velocity to keep fuel consumption down.

When the orbital airship reaches Mach three, it enters the low end of the hypersonic phase of the flight. This is where the vehicle has been tuned to fly. It's where it really shines. The airship is now a hypersonic waverider. A hypersonic waverider is a vehicle that takes advantage of the hypersonic wave. When traveling at hypersonic speeds, large shockwaves form around the airship. The shape of the wing is designed to ride on top of the shock wave. The orbital airship is the surfer of the upper atmosphere.

Hypersonic Aircraft

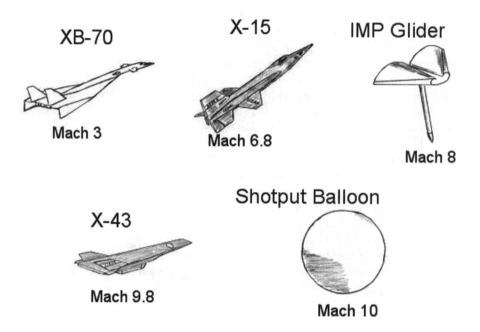

XB-70
Mach 3

X-15
Mach 6.8

IMP Glider
Mach 8

X-43
Mach 9.8

Shotput Balloon
Mach 10

Figure 11-1: Hypersonic Vehicles

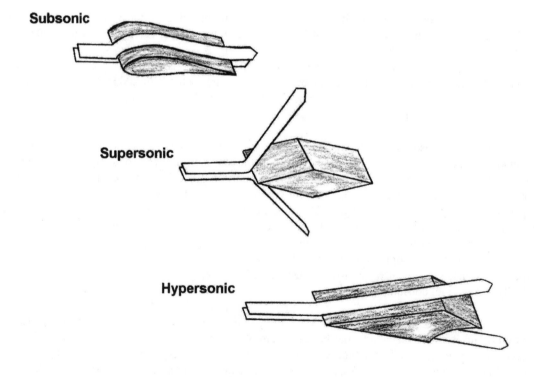

Subsonic

Supersonic

Hypersonic

Figure 11-2: Subsonic – Supersonic – Hypersonic Flow

To understand what a hypersonic waverider is, you need to know a little about high speed airflow. We'll start at the slow end with subsonic airflow. Air encountering a wing at subsonic speed will flow around the wing, with the airflow staying in close proximity to the surface (see Figure 11.2). As the wing moves fast, a wave of air builds up in front of it. Imagine a fishing boat on a calm lake. As the outboard motor roars, a bulge of water forms at the nose of the boat. The same bulge, although made of air instead of water, forms in front of the wing. The bulge increases as the speed increases, reaching a peak at the speed of sound. It is the bulge, or wave of air that is the sound barrier. This is the boundary between subsonic and supersonic flight. Flying a little faster, the wing passes though the wave. When aircraft are flying through the wave, they are said to be transonic. After they fly through the wave they are traveling supersonic. The air flowing around the wing acts differently on the other side of the sound barrier. Instead of flowing around the wing, it deflects off at steep angles.

As the speed increases the amount of air deflecting off the wing, the shockwave is pushed back close to the wing. There is no set speed where supersonic becomes hypersonic. Experts say that a vehicle is moving hypersonically when the shock wave is bent significantly back towards the wing.

Accelerating to Orbit

With wings for support, the propulsion system can run at lower thrust than a conventional rocket. The propulsion system is not lifting the dead weight of the vehicle. Air drag is the main obstacle to overcome. The propulsion system of the airship will determine how long it will take to reach orbit. The airship will need to accelerate to about 17,500 MPH. The two extremes are twelve hours on the fast side and five days on the long end. The shorter and higher the thrust is, the more efficient the vehicle. Efficiency is not the goal of ATO. A steam ship plying the Atlantic is very inefficient compared to a 747. Yet the steam ship is cheaper to operate and can carry more cargo.

Reentry

In returning to the Earth, the Ascender really shows its advantage over traditional spacecraft. The orbital airship can reenter the Earth's atmosphere without firing a retrorocket as conventional spacecraft do. Instead of firing retrorockets, the orbital airship throttles back the engine and points the nose up. By putting its massive lower surface into the air stream, orbital velocity is rapidly bled away and reentry commences. The amount of drag on the airship can be greatly varied. The maximum drag would occur when the nose of the airship is straight up. The entire bottom of the airship would become the world's largest speed brake. When the nose is pointed in the direction of travel, the drag is at its lowest. The airship can make very fine adjustments to pitch the nose during reentry. With these adjustments the reentry can be very precise.

There is enough atmosphere in orbit that the airship will need to keep the engines running to stay there. The options are to keep the engines running, climb to a much higher orbit, or just not stay at all. The solution that requires the least amount of energy is to just not stay. Drop and run, deploy any cargo and crew modules, and come home.

The ability to create high amounts of drag at high altitude allows the slowdown much sooner and at a higher altitude than traditional reentry vehicles. By loosing velocity before it reaches the lower thicker atmosphere, the reentry temperatures are radically lower. The airship can reenter without a heat shield. This makes reentry as safe as the climb to orbit.

Reentry for this spaceship doesn't end at the ground but at 140,000 feet. Instead of screaming in for a splashdown or landing on a runway, the orbital airship comes to a stop in the atmosphere, docking back with the Dark Sky Station. There it will be refueled, loaded with new cargo, and sent off again.

Anatomy of a Vehicle: Orbital Airship

The orbital airship is the largest aircraft ever conceived. The density of the surrounding medium is the driving factor in getting anything to float. It's easier to float objects in water than air because water has greater density. For the same reason, it's easier to float a balloon at sea level than at 100,000 feet. To lift its weight to 180,000 feet, the orbital airship needs to be enormous. Current estimates project the orbital airship to be 6,000 feet long, or just over a mile.

The "vee" shape was chosen for its aerodynamic characteristics of stability and low drag at high altitudes. A conventional blimp shape has lower drag only when it's acting like a blimp. When it tries to act like an airplane, the drag leaps up. It's as if the blimp has one very big stubby wing. The orbital airship needs to be a blimp and an airplane.

The shape must also be stable over a huge range of velocities. The orbital airship must be stable when barely moving, through moderate subsonic speeds, while penetrating transonic and supersonic velocities, and then all the way to hypersonic orbital velocity.

The vee wings are not entirely flat. The nose of the airship is set at a slight angle up relative to the rest of the vehicle. This modifies the hypersonic wave and allows most of the vee wing to fly at a flatter angle, reducing drag.

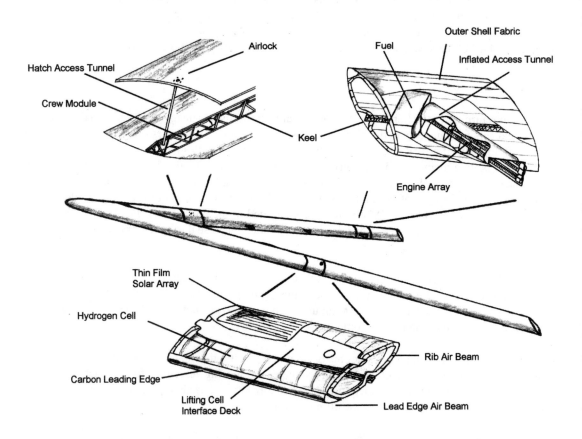

Figure 11-3: Orbital Airship

There are many phenomena in aviation that are described as barriers. The "sound barrier" is the classic. In the 1950's, many experienced engineers were convinced that the sound barrier could not be broken. At the time, there was even an impressive amount of data to back them up. Today there is the L/D barrier for supersonic aircraft. L/D refers to the ratio of lift provided by the wings to the total drag of the aircraft. This ratio has become an indicator of how efficient an aircraft is. Sailplanes have L/D ratios of over 60 to one. The Space Shuttle has a L/D ration of just over one to one. The L/D barrier for hypersonic aircraft is thought to be around five.

The orbital airship will need a, L/D ratio much high than that. It is important to note that the "L/D Barrier" was calculated for conical, or cone-shaped vehicles. We simply don't have the data to correctly model other shapes of hypersonic vehicles. Early tests indicate that lift-to-drag ratios much higher than those needed for ATO are possible.

Envelopes

In the simplest terms, the orbital airship is a big, fast balloon. However, it's really dozens of balloons all flying together. These small balloons are all contained within a single large balloon. The inflated structures—a fancy way of saying the balloons—consist of an outer envelope (the large balloon), the inner cells (the small balloons), and inflated trusses.

Structure

The structure consists of four primary components: an outer pressure hull, a rip-stop polyethylene envelope, a Kevlar upper shell, and the keel.

Outer Envelope

The basic vee shape makes up the hull's outer envelope. The outer envelopes will utilize nylon or rip-stop polyethylene, or another similar material chosen for its resistance to UV damage, low gas permeability, and other properties necessary for the unique environment of near space. It maintains its shape through pressure. Nitrogen gas fills the space between the outer envelope and the inner cells. As the gas expands when climbing, the nitrogen is vented overboard. This prevents the airship from "popping." When coming back from orbit, the gas contracts. To keep the airship from deflating, nitrogen is put back. Nitrogen is carried onboard in liquid form and flashed back into a gas and then released into the outer envelope.

Inner Cells

The hydrogen is contained inside an array of lifting cells. These cells are in rows inside the primary envelope. The inner lifting cells will be made of a nonmetallic-coated Mylar, chosen primarily for its low gas permeability. We will incorporate the ability for gas transfer between the lift cells and the ability to replenish lost gas. When initially filled, there is only a small amount of gas in the individual cell. It has the look of a jellyfish, a bubble of gas at the top with long sheets of cell material hanging down. As the airship climbs the gas expands. It is the same effect as when your potato chip bag puffs out when you go to the mountains. When the airship reaches maximum buoyancy altitude, the cell is completely inflated.

There will be up to 20 lift bags per arm. Multiple cells prevent "sloshing" of the lifting gas. If the hydrogen were allowed to move freely inside the length of the wing, the airship would rip itself apart. When the nose of the airship pitched upward, as it does when climbing, all the hydrogen would rush forward. There are two problems with this. First, the airship will violently pitch upward. The second and worse problem is that all that hydrogen, which is providing many tons of force carrying the airship, wants to apply that force to the inside of the nose of the vehicle. In an eyeblink, the nose would fail, causing the airship to come apart.

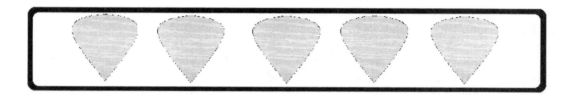

Figure 11-4: A row of inner lifting cells inside an outer envelope

Early Navy submarines had the same problem. These submarines had large ballast tanks that ran along the sides of the hull. These tanks were filled with air when the submarine needed to surface. These tanks acted like the inner cells of the airship. The problems arose when the tanks were only partially full. This happened every time the submarine was just starting on its way up or down. If a wave pushed the nose up just a few degrees during this process, all the air in the tank would rush to the front. All the water would rush to the back. The submarine would lose control and sometimes sink. After a significant number of lives and submarines were lost, designers divided up the ballast tanks into many individual cells. This change eliminated the problem entirely.

In many ways, the dynamics of an airship are closer to submarines than to other aircraft.

One of the most common questions asked about ATO is about meteorites. "What happens if a meteor popped the airship?" The answer is very little would happen. A balloon pops because the air on the inside is at a higher pressure than the air on the outside. The inner cells of the airship are "zero-pressure" balloons. The pressure on the inside of the balloon is the same as the pressure on the outside. There is no difference in pressure to create a bursting force. All a meteorite would do is make a hole. The gas would leak out staggeringly slowly. On the high altitude airship Ascender 175, built in 2004, there were six fourteen-inch diameter vents along the top of the envelopes. If all six vents were opened at once, it would take over twenty minutes before the airship would slowly start to descend.

A meteorite would likely make a smaller hole. The odds are that the meteor would not strike directly on top of the airship. A hole on the side would leak even slower than a hole on the top. All three vehicles in the system will have the ability to pump gas out of a damaged cell and replace it in flight. With the multiple cell system, the loss of a single cell would not mean the end of a flight. Unless a meteor hits a person or a vital piece of equipment, it is more of a maintenance issue rather than an emergency. Even the heaviest meteor showers would pose little risk. A meteor shower mission would actually be a great research opportunity.

The vehicle is so large that conventional thinking about the structure is inadequate. Each arm of the vee must be stable independently of the other. It is as if the two arms were flying in formation with each other. The arms can be thought of as mile-long levers, both acting on the apex of the "V." It is better to control those loads through flight dynamics than to add the significant structural weight to brute-force them together.

Inflated Structures

Much of the internal structure will be made out of balloons. The word "balloon" doesn't sound very structural. The industry that designs and uses them has assigned names like "inflated structure" and "air beam" so people will feel better about it. If the term "inflated structure" brings bouncy castles at a child's birthday party to mind, you're not alone. However, balloons as a building material have come a long way. Airplane wings, bridges, and antenna towers have all been made with just fabric and air pressure. Even everyday items are getting inflated structures. There is even a camping tent that uses an air beam in the place of poles.

In 1959, the Goodyear tire company built an inflatable airplane for the army. The wings, tail, fuselage, everything but the engine, was blown up with air. They called it the Inflato-Plane. The idea was to drop an Inflato-Plane to pilots who had been shot down over enemy lines. They would blow up the original plane and fly to safety in the Inflato-Plane. The plane flew terrifically. It had a speed of 72 MPH and could climb to 10,000 feet. There was even a two-seat version built for training. Sadly, the Army did not go forward with the project, and Goodyear dropped it.

Figure 11-5: Goodyear Inflato-Plane

Figure 11-6: Air Beam Tent
(Courtesy NEMO Equipment Company)

Figure 11-7: Inflated Beam
(Courtesy Vertigo Inc.)

If you're still not convinced about the strength of inflated structures, think about being run over by a car. It is the inflated structure, the tire, that gets you.

Hard Structures

A Kevlar deck runs along the entire length of the airship. This deck acts as the upper part of the lifting cells. The plastic of the cells is directly sealed to the edges of the upper deck. The deck provides all the plumbing services required for the cell. The vents, hydrogen pumps, and fill valves all connect to the lifting cell through the hard carbon deck. This eliminates the need to attach anything to the delicate film of the lifting cell.

Pumps and conduits are imbedded in the deck. Hydrogen is moved between the cells through this ductwork. This is the attitude control system. Hydrogen pumped forward and aft makes the nose go up and down. Hydrogen pumped from the one wing to the other causes the airship to roll from side to side. Venting allows the hydrogen to be dumped outside, reducing the overall lift of the airship.

The main internal structure is the keel. Like the keel of a boat, the airship's keel runs the entire length of the vehicle. The keel is made up of both inflated and rigid trusses. A truss-type structure of carbon composite will make up the primary underlying internal frame of the vehicle. This structure provides rigidity for the least amount of weight. Inflated ribs run along the inside of the leading and trailing edges of the wings. This maintains the wings' airfoil shape.

The leading edge of the outer envelope needs to be stiff. Any flexing will alter the hypersonic flow over the wing and greatly decrease performance. It will bear much of the aerodynamic load. A carbon fiber strip will stabilize the lead edge and give it strength. The strip will run from nose to tail all along the "V." This is a common feature of blimps, although instead of a leading edge, they have a nose in the front. An aluminum cap stiffens the nose. The leading edge cap also provides mounting for the electrical drag reduction systems.

Control

Airplanes use rudders, elevators, and ailerons to give the pilot control. These control surfaces deflect air or change the shape of the wing. As airplanes climb higher and higher, the control surfaces become less and less effective. At higher altitude there is less and less air to work with. As the air gets thinner, the control surface needs to get bigger to have the same effect. At an altitude above 100,000 feet, control surfaces need to be ridiculously huge. These vehicles need a more practical method of control.

Airplanes that fly to extreme altitudes need to use small rocket thrusters for control. Both the NASA X-15 rocketplane and Scaled Composites Spaceship One have thrusters to take over when their traditional control surfaces stop working.

Due to its great size, the atmosphere has an impact on the airship all the way to orbit. In the thin-air high altitude environment, neither the control surfaces of conventional aircraft nor thrusters can do the job of controlling the vehicle. The atmospheric forces needed to be overcome by these control systems are just too strong—no rudders, no elevators, no thrusters. At 250,000 feet, the size rudder needed to turn a mile-long airship would be impossibly large.

The orbital airship will have no moving control surfaces. To have control, the orbital airship will use gas-shifting and vectored thrust. The ability to rotate the airship to any orientation about its axis makes the entire airship a control surface.

These control surfaces are very effective at high speed. As the airship slows down, they get less effective. Large ocean-going ships have the same problem. The slowest speed they can go and still control the ship is referred to as the minimum speed required to maintain headway. Big ships can't just barrel into port. Yet at port is when they need the most maneuverability. Enter the tugboat. The tugboat pushes and pulls the larger ship, providing the maneuverability it needs.

The orbital airship has the same problem as the large ship. The orbital airship doesn't have the slow speed maneuverability needed to dock with the Dark Sky Station. Motors and propellers could be added, but for a high weight penalty. The better solution is the same as for the ocean-going ship—tugboats. The first stage airship can act as a tugboat. It can assist the larger vehicle move into the docking position.

Propulsion

No matter how you do it, getting to orbit is about velocity. The orbital airship needs to accelerate to 17,500 MPH. An electric propulsion system will provide the thrust. Electric engines sound like the stuff of science fiction. They may not power the evil empire's space fights, but they have been powering various real spacecraft for over three decades. The best example of the use of an electric engine is its use in the NASA Deep Space One probe. Other space probes are shot on their one course for their one mission. The electric engine on Deep Space One allowed the probe to zip around the solar system, accomplishing several missions. Electric engines put out very low thrust and use almost no fuel. It uses electricity to accelerate the propellant, providing the thrust. The drawback of electric engines is that they use tremendous amounts of electricity.

The apparent disadvantage of the low thrust of an electric engine becomes an advantage for the orbital airship. With low thrust comes long duration. Some electric engines have run as long as a year! Electric propulsion systems have taken huge strides in the last few years. The orbital airship has several options currently to choose from.

Traditional electric propulsion engines such as Ion, Hall Effect, and Arc Jet do not have sufficient thrust to drive the orbital airship. However, huge advances have been recently made. In 2006 alone the European Space Agency announced a new ion engine with four times the performance of anything previously built. NASA, not to be outdone, announced a short time later that they had developed an ion engine that had ten times the performance. At the current rate of progress, in ten years a pure electric thruster may be available to do the job. Until then, the combination and hybrid systems are the stronger candidates.

The orbital Ascender's engine will be a merger of a chemical rocket engine and an electric rocket engine, although there are a lot of candidate engines out there.

VASIMR

The VASIMR (Variable Specific Impulse Magnetoplasma Rocket), engine is one of the most advanced electric based thrusters. There are three elements of a VASIMR engine. The first is the ionizer. Gas entering the engine—hydrogen has been used in testing—gets its electrons stripped off. The gas molecules are now ions. Ion simply means the molecules have an electrical charge. The ions are then heated with radio waves. By heating the ions, up to millions of degrees Kelvin, the total energy is increased. The high energy ions are then released into a magnetic rocket nozzle. The nozzle converts the energy of the ions into forward thrust. The engine can be throttled, either adjusted down to a high efficient low thrust output or cranked up to a low efficiency high thrust output.

An operational version of the VASIMR engine was proposed for the International Space Station. The engine would be used to maintain the orbit of the space station. The station continuously drifts downward from its orbit, drag from the upper atmosphere being the culprit. Without periodically being boosted back, the station would fall back to Earth. Currently, the Russian capsules and American Space Shuttle that dock with the station provide that boost.

The engine unit was low-cost and self-contained. It stored electrical power with batteries during off-peak usage time from the station's solar panels. Unfortunately, the proposal was not accepted. The initial design study for that system does provide an excellent baseline for evaluating flight weight, as opposed to lab weight, VASIMR-based systems.

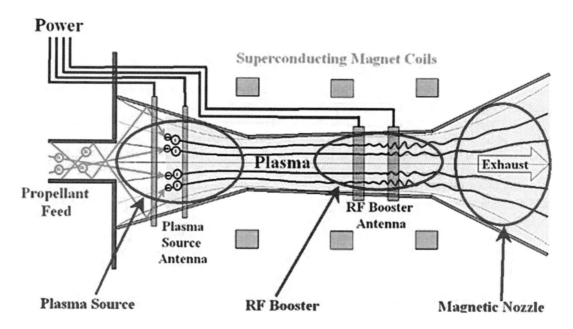

Figure 11-8: VASIMR Engine (Courtesy Ad Astra Rocket Company)

Electrically-Controlled Solids

Solid fuel rockets have been the workhorses of space travel. Nearly every space system will use them in some way. Small solid fuel rockets, like the Scout, have placed dozens of satellites into orbit. Large solid rocket motors make it possible for the Space Shuttle to fly. Solid fuel rocket motors have sufficient total energy to drive a small airship to orbit. The problem is the rate of energy release. Put on one of the Space Shuttle's side booster motors, and it would tear right through it. There are slow burning rocket fuels out there; however, the price they pay for burning slow is low efficiency.

There is a new class of solid rocket just developed in the past few years—solid rocket motors where the level of thrust is controlled by an electric current. A small charge is applied to the surface of the fuel. The fuel is formulated so it will only burn while the electrical power is on.

There are several advantages of electrically-controlled solids in their simplicity and their low electrical power need. The biggest appeal of the electrically-controlled solid is its ability to throttle to low thrust without loosing efficiency. It does not have the efficiency of the other candidate propulsion systems. The other drawback is in the direction the market is taking this technology. Electrically-controlled solids have found a market niche as microthrusters in very small satellites. These motors are no more than an inch long and can provide very precise thrust, although on an extremely small scale. This is wonderful for satellites the size of a soda can, but not useful for giant airships. If the development of large-scale motors is restarted, these motors could be an excellent solution for the smaller developmental ATO vehicles.

Hybrid chemical/electric

The newest engine candidate is a combination of six motor types rolled into one. The name "the liquid/solid/pumped plasma, magnetically accelerated, radio frequency (RF) aerospike engine" is just way too long. Even the acronym-loving space program would find it a tortured creation.

For clarity it is called the "symphony" engine. The basic concept of the symphony engine is accelerating fire. At the core of the symphony is a hybrid rocket engine. The propellant of a hybrid engine is part liquid and part solid. The fuel is solid, sometimes even hard plastic. The oxidizer is a liquid. Nitrous oxide or liquid oxygen are the most common. The hybrid engine is very safe. The chances of the fuel exploding on its own is the same as the odds of your plastic pen spontaneously exploding in your hand. This inherent safety was one of the main reasons this type of engine was selected for the first private spacecraft, Scaled Composite's Spaceshipone. Combined with the hybrid core are RF heaters, magnetic nozzles, a linear accelerator, and a plasma turbo pump.

RF heaters are radio antennas that use radio waves to add heat to the engine combustion chamber. This is similar to a microwave oven. One of the chief problems with a hybrid rocket engine is slivering. The solid fuel of the hybrid does not burn evenly across the surface. Small chunks called slivers break away and burn. These slivers burn suddenly causing an increased in pressure in the engine. The slivers are very small, the largest being the size of a grain of sand. This chunky burning makes the rocket engine very rough. Hybrid rocket engines have the highest vibration loads of any type of rocket engine. The RF heaters on the symphony engine vaporize the surface of the fuel. This eliminates slivering and burns the fuel more efficiently.

RF heaters in the needed frequency range were once the domain of experimental fusion reactors. They were expensive, classified, and nearly impossible to get. In a dramatic shift, RF heaters are now used mainly in bulk commercial applications. They are extensively used in the textile industry. RF heaters are the high speed dryers for your blue jeans and dyed fabrics. The commercial use of RF heaters has caused the price to drop and the efficiencies to rise.

Key elements of the symphony engine have been tested. A sounding rocket with a small prototype symphony engine called a "quad" is under construction. An initial flight test series will begin in 2008. The quad engines will be used on small test airships. Their firing times will initially be less than a minute. This will be gradually extended to over five days as the program progresses. The first full symphony engine test program will begin in the fall of 2009.

Crew

The crews live and work in pressurized modules. These modules will be standardized throughout the entire ATO system, the same modules used on the Dark Sky Station and the first stage airship. They will be cylinders made of carbon fiber, foam, and Kevlar.

Flying the orbital airship resembles operating a ship at sea rather then flying an airplane. For example, to turn the airship around, planning and checklists will take over an hour to do.

With the ability to conduct repairs in flight comes the need for the technicians and engineers to do the repairs. The orbital airship will have inflatable access tunnels that lead to pressure modules. These modules will be located in such an order to provide access to in-flight maintainable areas. Maintainable systems would include engines, power generation, and life support systems.

Wireless networks will be used to distribute avionics over the structure to remove the need for wiring. This will facilitate further developments since there won't be a need to change wiring between nodes when changing functions. This system has been operated successfully and reliably at high altitudes in existing vehicles.

Power System

The electrical power generation system will be the heart of the orbital airship. The power system selection is directly determined by the propulsion system used and therefore the time it takes to reach orbit. A purely electric-based system, such as a traditional ion engine, requires a tremendous amount of power, but it allows for a slower, more efficient climb to orbit. A purely chemical engine requires little power, but the fast burns and resulting high loads make the airship too heavy. A power trade study was conducted looking at standard power supply type vs. time. The study showed that flights shorter than five days should be run on batteries. For climbs to orbit taking five to ten days, fuel cells are the best option. Longer duration climbs would require solar panels.

Solar

Photovoltaic arrays (solar panels) will convert sunlight into electricity. The upper surface of the orbital airship provides a huge area for solar panels. Solar panels can now be made as thin films. These solar panels feel like plastic wrap and are extremely light when compared to the glass plate-style cells of just a few years ago. Not all of the solar panels can be in direct sunlight at any given time. The angle of the sun and orientation of the airship has an impact on electrical production. The working assumption is that only 40 percent of the panels are exposed to sunlight at any given time.

Batteries

Battery technology made a giant leap forward with the introduction of the commercial lithium ion battery by Sony in 1991. Great advancements in battery technology continue at a very fast pace. Every year breakthroughs are increasing the power densities of batteries. If the current progress continues, when the first orbital airship tests are conducted they could be powered completely by batteries. Here are some potential battery technologies:

Nanophosphate lithium ion are reaching 3,000 watts per kg. For comparison, NiCd batteries have an energy density of 35 watts/kg.

Ceramic ultra-capacitors promise power densities of up to 2.5 times that of conventional lithium ion batteries. They can provide up to 4,000 watts per kg. Carbon nanotube ultra-capacitors are being investigated at the Massachusetts Institute of Technology (MIT). The researchers at MIT are projecting energy densities of 100,000 watts per kg.

For raw power storage, nothing can beat a Maglev flywheel. The volume inside the airship allows for flywheels of a truly tremendous size. With a large diameter of flywheel, the mass density can be lower. Experimental city buses are now being run on these flywheels. More study is needed to see if this is a practical power source.

Micro-nuclear batteries are moving very rapidly from research to production. These batteries are real power generators. They work much like solar panels, employing the photoelectric effect instead of using the electrovoltaic effect. However, instead of the sun, they use a microscopic amount of tritium gas. The gas is imbedded in pits only four ten-thousandths of an inch wide. The battery is made using processes similar to a computer chip. Currently this technology is being developed for cell phone use. It will be several years before this technology could be scaled up for orbital airship use. Yet that may be just when it's needed.

Catastrophic failures vs. evaluate and repair

The slower pace of the orbital airship's climb to orbit introduces another new concept: safe space travel. The single most dangerous aspect of space flight is the rocket ride up. Coming in a close second in the danger race is the re-entry into the atmosphere on the way back down. During the launch of a conventional rocket, the failure of any one of a million components can cause a catastrophic failure. Abort decisions are required to be made in seconds or less. There are precious few failure procedures that allow for the continuation of the mission. In those few minutes of a rocket launch, so much energy is released in so short a time that perfection is not a goal but a requirement. Most aborts for crewed missions revolve around efforts to save the crew. Slowing down the pace and stretching out the energy expended from four minutes to 20 hours fundamentally changes the nature of emergencies in a vee ship. Instead of an extreme situation where every second counts, nearly all problems can be handled with an "evaluate and repair" strategy.

Compare an electrical failure on the Space Shuttle vs. one on an orbital airship. A primary electrical failure requires an immediate abort of the mission. Depending on where the Space Shuttle is in its flight, this could involve separating the Shuttle from its large fuel tank, jettisoning the side booster rockets, or even bailout by the crew. In any event, a landing site must be determined and a glide path selected because in a few minutes, one way or the other the Space Shuttle is going to be on the ground.

For the orbital airship, this emergency gets downgraded to a problem. Even without electrical power, the vehicle still glides forward, stable and safe. The crew would notify the ground and engineers would begin evaluating the problem. There are hours, not seconds, to correct the problem. The crew could inspect and repair the fault themselves and then continue with the climb to orbit. If the problem cannot be corrected, the vee ship would slowly glide back down over several hours.

Even more fundamental, if the massive rocket fuel tanks are not there, they cannot explode. Traveling to Earth's orbit will be safer than flying in an airliner.

The development process

The orbital airship not only requires many new technologies, but also requires them to work together. The way to achieve this is through complete, incremental vehicle development.

In this method you bring together all the components, regardless of their stage in development, and put them in a single vehicle and fly them. There is a bigger challenge than the individual new technological elements. It's integration—getting them all to work together. Complete vehicle incremental development lets you tackle integration early in the process.

Mach Gliders

There are still many questions about hypersonic airship flight. Questions like: what range of skin material flexibility is tolerable? What is the maximum glide ratio possible? What is needed for a low cost research tool to answer those and other questions? That tool is the Mach Glider. Mach Gliders are miniature versions of the orbital airship.

These mini airships are the real research workhorse of the airship to orbit program. The Mach Glider is made of thin film plastic such as Mylar or polyethylene. Mach Gliders can range from 10 to 40 feet in length.

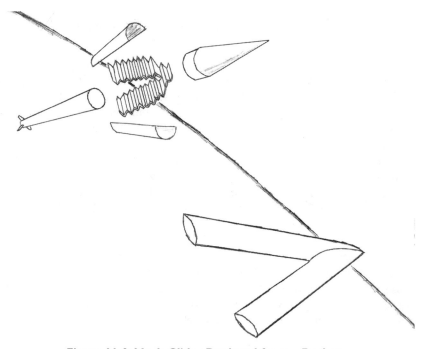

Figure 11-9: Mach Glider Deployed from a Rocket

Mach gliders will be deployed from rockets and airships. When ejected from a rocket at altitudes above 100,000 feet, the Mach Glider will glide down at velocities above Mach one due to the low density of the atmosphere at that altitude. The first Mach gliders will be simple vee-shaped balloons. They will be deployed just like the Micrometeor Paraglider from 1964. For advanced research, the Mach Glider can have a rocket motor on board to propel it to higher altitudes and velocities. Mach gliders can cheaply explore the entire hypersonic flight envelope.

High altitude airships are perfect launch vehicles for Mach gliders launched from 60,000 to 100,000 feet. After the airship releases the glider, the glider can act as a communications relay. This can reduce the weight of the telecommunication system on the glider. The airship can also take video of the glider's flight. The airship can also be used to pinpoint the glider's landing site and aid in the recovery operation.

Before launch, the Mach Glider is folded and placed inside a split hollow foam cylinder. The cylinder is slid inside the airframe of the rocket. When the rocket reaches peak altitude, the cylinder is ejected from the rocket along with an inflation unit. The inflation unit is a small gas cylinder with a valve and control circuit. Upon ejection from the rocket, the halves of the foam cylinder drop away. The inflation unit fills the Mach Glider with gas, causing the Mach Glider to take on the "V" aircraft shape. After inflation of the aircraft, the inflation unit drops away. The Mach Glider then proceeds on to its mission activities. The Mach Glider will operate, in duration, from minutes to several hours depending on the mission requirements. At the end of its flight, the Mach Glider will be burst and will drift down and can be recovered by ground personnel.

Many of the tools needed to fly Mach gliders are already complete. The sounding rocket to be used has flown several times. On one flight, it ejected a payload similar to the way it would eject a

Mach glider. An inexpensive high altitude airship is needed for several of the tests. As I write these words I'm sitting next to such a vehicle.

Mach Glider program outline:
- Cylindrical balloon construction and indoor rapid inflation tests (completed)
- From sabot (carrying sleeve for the cylindrical balloon) ground deployment (completed)
- Low altitude, 10,000 feet, rocket deployment and inflation of cylindrical balloon
- Mach Glider construction and indoor rapid inflation tests
- Rapid inflation tests from sabot
- High altitude inflation and drop from airship
- 160,000 foot Mach Glider drop from balloon
- Rocket-deployed Mach Glider at 200,000 feet
- Rocket-deployed Mach Glider at 200,000 feet with a rocket booster on the Mach Glider
- Rocket-deployed Mach Glider at 400,000 feet
- Rocket-deployed Mach Glider at 400,000 feet with a rocket booster accelerating the Mach Glider to hypersonic velocity.

Anatomy of a vehicle: The Block 4 Mach Glider

The Block Four Mach Glider will fly a mission similar to the IMP Glider in 1964. Both are inflated gliders, both deploy from a rocket at 400,000 feet, and both have Mach five glide speeds. The Block Four Mach Glider will depart from its forebear; during the Mach five glide, it will fire rocket engines.

The Block Four Mach Glider is 30 feet long and 20 feet wide at the wingtips. It will be filled with helium, but not enough to float. In the center of the vee will be an early prototype of the symphony engine. The goal of the Block Four Mach Glider is to sustain level flight at Mach five for 30 seconds. The run will start when it is completely inflated and stabilized at 380,000 feet. The glider will level off at 260,000 feet and engage the engines. After 30 seconds of powered flight, the vehicle will continue reentry and glide back to Earth. The Mach gliders will evolve into crewed vehicles—the Transatmospheric Airship.

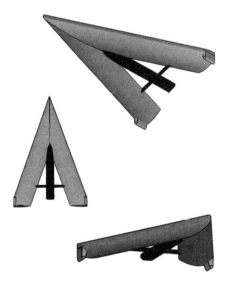

Figure 11-10: Mach Glider Block 4

Transatmospheric Airship

Author's personal bias: this is the vehicle I'm looking forward to building. The Transatmospheric Airship is the halfway vehicle. It is a suborbital test bed. A series of these airships will need to be built. They will be designed to reach space, at 62 miles, but just at velocities below that required for orbit. It will be the suborbital test bed for ATO technologies. The Transatmospheric Airship will be a crewed vehicle. On board will be a pilot and an engineer. At approximately 2000 feet long, it will be too large and light to survive long in the lower atmosphere. These will be the first vehicles to be assembled at the Dark Sky Station.

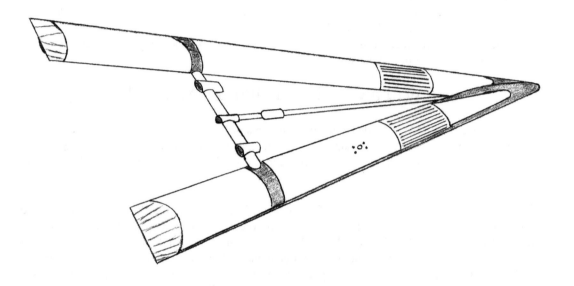

Figure 11-11: Transatmospheric Airship

There are two Transatmospheric airships in the planning stage. Mach gliders will lead the way, laying out the aerodynamics and basic operating parameters.

Block One climb and hold altitude

The first vehicle will have modest goals. It will climb from 140,000 feet to 200,000 feet and hold altitude for a short duration. 200,000 feet will be above the altitude it can reach by floating alone. It will require the use of its engines and wings to reach and stay at 200,000 feet.

Block Two suborbital flight

The Block Two vehicle will be large enough to accommodate the powerful engines required to take this airship to space. It will climb 62 miles to space but not go to orbit. Once the Block Two Transatmospheric airship reaches peak altitude, it will glide back to the Dark Sky Station. It may be possible to upgrade the Block One vehicle into the Block Two vehicle. This would accelerate the development process. The Block Two vehicle has commercial potential on its own. It can be used for suborbital space tourism and high speed package delivery. The Transatmospheric airship would be the ultimate atmospheric research vessel.

Figure 11-12: Transatmospheric Airship heading for space

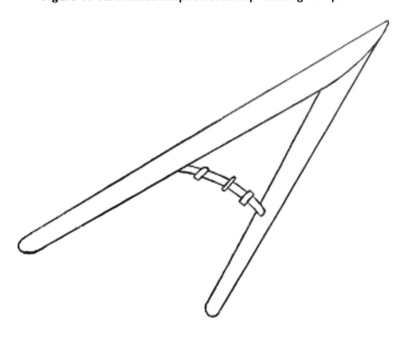

Figure 11-13: Small Orbital Airship

Small Orbital Airship

The first orbital airship will be a direct scale-up of the Block Two Transatmospheric airship. The big difference between this vehicle and the previous one, besides size, will be electrical power generation. This airship will carry the power system necessary to reach orbit. The operations of the Block

Two Transatmospheric airship and the initial orbital airship will show the fundamental truth about this technology. There is no major difference between a suborbital and an orbital airship. There is no solid barrier, no giant boost needed, no need for an exotic warp drive. The difference is the incremental increase in power—nothing more, nothing less. All of the issues of going to orbit and returning can be worked out on the suborbital vehicles.

Large Orbital Airship

Orbital airship development must be in sync with high altitude facilities development. The orbital airship requires a large floating construction facility. Before the orbital airship can be built, the Block Three Dark Sky Station must be in place. This airship is the big guy. The large orbital airship will be 6,000 feet long and capable of carrying 40,000 pounds to orbit.

Figure 11-14: 6,000 Foot-Long Orbital Launch Vehicle

Chapter 12
A Walk Down the Hall

Aubrey looks out a window and sees a spacesuited figure walk to the edge of the station and leap off. For a moment, he's stunned; then, in panic, he races down the corridor that opens up into a large auditorium. Aubrey scans the banner above the monitor and lets out a deep breath as he reads "2nd annual Strato-diving Championship." The crazies can have their fun, he thinks, but where was the supersonic paper airplane drop contest being held? He wants to throw a paper airplane twenty-four miles. "Before you throw yourself off, remember to leave me on the table," comes the tweety voice of iPal. "Just great," Aubrey sighs, "he's logged into the station's schedule system."

Switching his iPal to silent mode, Aubrey just stands and stares for the 30 minutes it takes Molly Myers, the current Strato Diving champion, to get sufficiently low enough to deploy her parachute. A camera mounted on her helmet shows her view of the plummet to Earth. "I'll keep my feet planted on terra firma," Aubrey thinks as the crowd cheered for Molly. Aubrey never saw the irony of his thought as he walks along decking that was suspended twenty-four miles above the Earth.

"Exactly what I was looking for," thinks Aubrey as he spies a map and directory on the wall, "Just like at the downtown mall." His eyes gaze down the list of labs, quarters, and arm access airlock staging areas—"whatever that is." Below is a list of events. At the top of the list is "Station Walk: A three-hour guided tour of the Dark Sky Station." "Three-hour walk," he thinks in disbelief. "The inside of the station is no bigger than a few dozen mobile homes, where is there to walk for three hours?" He reads the rest of the announcement, "Wear comfortable shoes. Tours start daily at 1:00pm at The View Lounge."

Aubrey looks at his watch. It's 1:05 p.m. He gives the map a glance. "Where did they put the lounge on this thing? Ok, back down the corridor, through the pressure lock, make a very sharp left, just past the restaurant."

As Aubrey approaches, a small group has gathered around a crewman wearing a blue jumpsuit. Aubrey can tell they were just ending the introduction to the station. "Well, that wraps up the history of the place. Counting the eighteen months of construction, this facility has been airborne for eight years now. Our first stop will be the bio lab."

Down another hall and they came to a hatch marked "Research Module Authorized Personnel Only." Their guide explained: "The labs need a lot of access to the outside. That means lots of ways to accidentally depressurize the module. The scientists get a little touchy about that, so we limit access. Also, it's a long way to the supply store, so the labs tend to share equipment even for widely divergent disciplines. For that reason and the depressurization issue, we keep them all together.

"We've identified thirty-two different types of bacteria living at station altitude. The food source is completely adequate for their needs, but we haven't been able to prove that they actually consume it."

"Excuse me, Professor," Aubrey recognizes the questioner as a passenger from the ride up, "What food source?" The scientist gives him one of those "don't they teach you anything in school" looks, then catches himself and smiles. "Millions of tons of organic material fall to the Earth from space everyday—amino acids, poly carbon chains, everything a growing bacteria needs. We just can't get the little buggers to eat it in the lab."

"Maybe it's the decor," comes a murmur from someone in the group. Even the biologist smiles. The anonymous mutterer has a point. The lab consists of inside-out gloves sticking into glass cabinets on one side and banks of computers on the other side. They are separated by a walkway wide enough for one and a half people. There is barely room for both the tour group and air molecules.

The hatch opens, and a woman in her early twenties stands there, reluctant to add one more person to the already-packed lab. "I want to introduce you to Sue. She's an exobiologist destined to find life on Mars. The extreme life we find up here is an excellent training ground for her."

"Whatever you say, Boss."

That is their cue to leave. The guide ushers the group to their next stop.

A few steps down the corridor and the tour group stops in front of a large airlock. The guide turns and gives the crowd a serious look, "We're now going to leave the safety of the habitat and enter the station's lifting arms."

The guide continues, "There is a ten-foot diameter inflated tunnel that runs the length of the arm. The atmosphere in the tunnel is at the same pressure as the rest of the station. Every 100 feet we'll be passing through a hatch. The hatches keep the entire tunnel from deflating if one segment should fail." A man from the back pipes up, "The tube we're in could pop?" The guide jumped in as a few in the crowd starting looking a little pale. "There is an inner and outer tube. Both would need to fail for the tube to depressurize. Each is made from Kevlar with a polyethylene coating. Even if you hacked at it with a knife, you would have a hard time getting through.

"There is an inner lifting cell swap out being done today. It will give us an excellent opportunity to see first hand what keeps us in the sky."

All along the tube, clear panels provide a view into the cavernous insides of the station arms. At each airlock they can see a mesh wall extending three hundred feet up to the cylindrical ceiling. The mesh divides the cavern into giant blue fabric rooms. In each room, just above their heads, is the biggest balloon any of them has ever seen. Aubrey guesses that they are at their destination when he looks up and the balloon is gone. In its place is a three hundred foot waterfall of clear plastic. He never thought plastic wrap could look so beautiful.

Watching an inner cell being replaced, Aubrey notices crew members in pressure suits carrying a ten-foot long rectangular box. The guide explains, "It's just like changing a giant ink cartridge in a printer. The redeployment of the balloon is automatic. The hydrogen from this cell was moved into the reserve volume of the two cells next to it. When the new cell is in place, the gas is pumped back in. This way the station doesn't tip or drop. Nobody feels a thing."

The tour continues down the tube as the crew loads a balloon cartridge. They continue on for nearly 30 minutes down the tube, passing airlock after airlock. Finally they come to the end of the arm. There is a small hatch in the flexible wall. A sign on the door shows a miner in a Star Trek uniform. The caption beneath proclaims, "Anti-Matter Mine." The guide looks a little embarrassed, "We don't really mine anti-matter. This is the anti-matter research lab. Small amounts of anti-matter briefly form in the upper atmosphere. It's a whole lot cheaper to study naturally-occurring anti-matter than it is to build a giant supercollider and make the stuff. The first antimatter studies on balloons were conducted over the Arctic in the mid-1990's. With the station, it's a case of bringing the researchers to the objects being researched."

On the way back down the arm, the group repeatedly has to step to the side to let a running crewmember pass. The guide apologizes, "Station staff take advantage of the maintenance arm tube for jogging."

After the tour, Aubrey made his way to the station's lounge. The station's biggest extravagance and the most popular attraction is a three-by-twenty foot window. The panorama of the black sky with the blue arc of the Earth shining below is stunning. "This is what your average person wants from space tourism," he thinks. "Not the high Gee-load and vomiting; people want to stand on the veranda with a drink in their hand, watching the big ball of the Earth roll silently beneath them."

Aubrey has to suppress an urge to point out the window and say, "Make it so."

Aubrey orders a blue soda and notices a man in his late 80's gazing out the window.

"I helped put man on the moon in the sixties but I never thought I'd make it here myself," he says to the air. Trying to make conversation, Aubrey asks, "Have you ever taken the run to orbit?" The old engineer takes a long pull on his beer. "Just once, can't take the zero-gee toilet. Give me a view and a beer and I'm good." Aubrey shifts a little uncomfortably at this; he wasn't too fond of the zero-gee toilet himself.

At that moment the engineer stood up, pointed at the window and said, "Make it so." He turned toward Aubrey and grinned, "I've always wanted to do that."

Chapter 13
Development Tools

When building a bridge, you need more then steel I-beams. You need an army of dump trucks, bulldozers, cranes, and barges to get the job done. While developing the Airship to Orbit system, specialized vehicles will also be needed. These vehicles will be the tools that will do the research, test the components, and show the way. They won't be part of the final system—you don't need the bulldozer after the bridge is built—but they are vital components of the system nonetheless.

Mini Platforms

To develop technology in the extreme environment of near space you need to go there, a lot. One of the best tools for spending time in near space is the High Rack. A High Rack is a miniature platform for testing equipment at the edge of space. It consists of a stack of foam shelves held together with a carbon fiber frame. It is shaped roughly like a diamond and stands four feet tall. The shelves are loaded with tracking equipment, radios, computers, and cameras. All types of balloon combinations can carry High Racks aloft. They have flown under groups of zero pressure plastic balloons, one to six clusters of rubber weather balloons, and single giant balloons of all types. These vehicles typically fly from 90,000 to 130,000 feet. When the mission is done, a command is sent to release the balloons, and the High Rack descends by parachute. Whenever a new pump for an airship needs testing or a telemetry system modification needs a good shakedown, a High Rack is pressed into service. They are the real workhorses of the ATO program.

Figure 13-1: Away 27 High Rack

Figure 13-2: High Rack in Flight   **Figure 13-3: View from a High Rack**

ML Sounding Rocket

The ML sounding rocket is designed to carry Mach Gliders and deploy them in the upper atmosphere. When fired from a small DSS, the rocket can carry its payload to 400,000 feet. The ML has been one of the early economic spinoffs from ATO development. "ML" stands for Micro Satellite Launcher. Performing rocket flights with the ML has been an excellent source of funds for the program.

The ML is twenty feet long in the two-stage configuration. The airframe and fins are made from carbon fiber and the nose cone is Kevlar fiber. The rocket deploys a parachute after performing its mission and is reusable. The rocket has flown on several motor configurations, from solid propellant to hybrid propellant. A new ML Rocket is being built that will incorporate an electric-based engine. The first flight is intended to dispel the myth that electric propulsion systems can only provide low thrust and can only run in a vacuum.

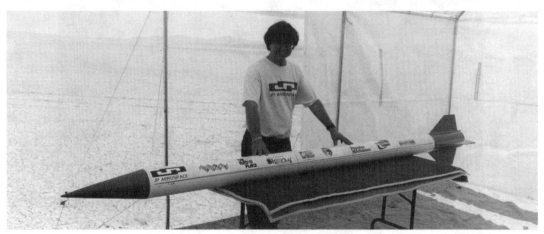

Figure 13-4: ML sounding rocket upper stage

Twin Balloon Airships

Building contractors have their backhoes and their cranes. Near space developers will have twin balloon airships.

Balloons are low cost and can lift heavy loads, but they have operational limits. High altitude airships have greater operational capabilities, but their large size equates out to large operational costs. A solution between the free balloon and the high altitude airship is needed.

That in-between vehicle is the twin balloon airship. The twin balloon airship is a small truss lifted up by two rubber weather balloons. Each balloon is mounted on a ring that allows forward and aft motion, but channels any side-to-side motion into the truss. To give the vehicle mobility, motors and propellers are mounted on the booms. By eliminating the big airship outer structure, the vehicle can be very light. In near space, light means it can be small. The first twin balloon airship, the Tandem, weights only 70 pounds. The entire structure is 35 feet long. It can fly to 110,000 feet and can remain aloft for eight hours. These crafts can be made larger and smaller to perform different tasks. The 35-foot standard Tandem and fifteen-foot Micro Tandem have been built. The 200-foot Tandem HL is under development.

The twin balloon airship will play a critical role in ATO development. It can drop Mach gliders, act as a communications relay platform, and perform the role of a camera and chase plane for first stage airship and Dark Sky Station flights.

The first Tandems will act as a test bed for new high altitude propellers and electrical propulsion systems. It will also be used as a training vehicle for new high altitude airship pilots.

Figure 13-5: Tandem Airship

Figure 13-6: Completed Tandem Airship

Figure 13-7: Eighteen-foot Scale Tandem Test Flight

Micro Tandem

Even test beds have test beds. The Micro Tandem is a fifteen-foot long version of the full-size Tandem. The full Tandem command/control system can be carried aloft and tested realistically. In addition, the Micro Tandem will be used as a training vehicle and for applications where only a very small payload capacity is required. This racer has a maximum speed of one knot with propellers. After the first two test flights with propellers, it will be fitted with the small development versions of the ATO's main engines. It will be the first airship to use chemical/electric rocket motors.

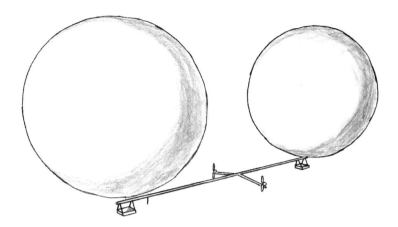

Figure 13-8: Micro Tandem

Tandem HL (Heavy Lift)

The Tandem HL will be the pickup truck, tractor, and crane of the Dark Sky Station and orbital airship construction. Its 200-foot long center truss will be a stressed film skin with an air beam core. Like the smaller Tandems, its balloon with be mounted on stabilizing rings. However, it will use large plastic balloons instead of latex. As a balloon climbs, the gas inside expands. In a latex balloon, the balloon stretches to accommodate the bigger ball of gas. Plastic balloons don't stretch. When plastic balloons are launched from the ground, they look like a giant ribbon of plastic wrap with a bubble of gas at the top. The extra plastic just hangs down until needed. On the Tandem HL, the extra plastic is rolled up on a carbon fiber drum. As the gas expands additional plastic balloon is unrolled. This keeps the ball of helium taut and stable.

The roller-deployed technique was demonstrated on the Away 28 High Rack flight. It is part of the common architecture and is used in Dark Sky Stations and orbital airships to replace damaged or worn inner lift cells.

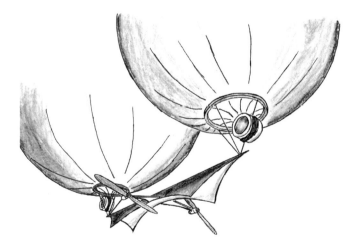

Figure 13-9: Tandem HL

After the Dark Sky Station is complete, the Tandem HL will be used as a general inspection and maintenance vehicle. It can even be used to ferry people and cargo between stations that are within a few miles of each other.

Jellyfish Balloons

Most of the data needed is about the atmosphere between 180,000 feet and 300,000 feet. This is higher than balloons can go today. Sounding rockets can spend a few precious moments there, but this is not enough. What is needed is long duration presence in that environment. A new type of balloon is being developed that can reach 300,000 feet and maybe beyond. The balloon that can reach these heights is the Jellyfish. The Jellyfish balloon is a balloon with a large open bottom. This layout increases the volume while decreasing the weight.

Over the North and South poles, there are tremendous upward currents of wind. These are caused by gravity waves striking the polar vortex. This may be what pushes the mesospheric ice clouds to their great heights. It's important to note that atmospheric gravity waves are similar to ocean waves, the difference being that the waves are in the air at the top of the atmosphere. This is not the same as the elusive gravity waves being sought by particle physicists. This opens the possibility for a balloon to float to space. A Jellyfish balloon could be flown to 200,000 feet in the polar vortex. Timed to ride the gravity wave, it could rise with the upwelling to 62 miles, officially where space begins, and above. The data that such a mission could provide is incalculable. Our knowledge of this region of our world would leap forward. Not only would the Jellyfish balloon be able to put instruments in this region for long duration, it would be interacting with the area. It is akin to diving into the ocean instead of studying it from the shore.

The interaction between gravity waves and the polar vortex is also suited for exploration by high altitude airships. Could an Ascender act like a sailplane, soaring the upper atmosphere like a mountain wave? Space soaring could become the ultimate aviation challenge.

MesoHab
Reseach Facility at 200,000 feet

Figure 13-10: Extreme Altitude Research Facility Based on the Jellyfish Balloon Concept

The configuration of the Jellyfish balloon was inspired by an accident. In September 1995, the JP Aerospace team was in the desert playing with balloons and rockets. The mission was a rocket launched from a stack of nine weather balloons. This was a low altitude test. The rocket was launched when the balloons had pulled it up to 300 feet. After the launch, the platform that contained the rocket released its balloons and came down by parachute. We didn't want the balloons drifting away, so they were attached with a 2000-foot tether line. The line was that long so it would not interfere with the free drifting of the platform during the launch.

After the launch, the entire team grabbed a hold of the tether line. It took over an hour to pull the sky anchors down. As each balloon came into reach, William Powell, the author's brother, would slash it, releasing the helium and easing the load. On one balloon, he cut the balloon horizontally above the nozzle. Instead of popping, the balloon took off. The top of the balloon was intact; however the bottom of the balloon was completely cut off. It looked like a giant flesh-colored jellyfish. The helium didn't flip it over or roll out the sides; it was completely stable. The sky jellyfish climbed at an amazing rate and was soon out of sight. I got out my notepad, knowing there was an important concept floating on that scrap of rubber.

Chapter 14
Floating to Space

Next morning, looking out the window, Aubrey hunts for the vehicle that will be the next step in his journey. The Earth and one of the massive station arms are the only things in sight. An idea strikes him, the lounge window. Two minutes of grabbing and stuffing, and he's packed and off. A small crowd has gathered by the big observation window. A two-mile long spacecraft that appears to stretch to the horizon now dwarfs the once huge station.

No time to gape—time to get on board. He makes his way to the lowest level of the station, following the "you are here" signs to get to the docking port. The orbital airship is docked under the station. This makes for direct access from the pressurized sections below; however, it puts the airship's docking port on top, 300 feet above the passenger module. Visions of a very long ladder fill Aubrey's mind. Heights always make him a touch queasy. The sign over a set of double doors lifts his anxiety: "Passenger Docking Elevators."

"World's first space elevator," iPal chimes in. "Only 21,999 miles to go." "You've got to admire those space elevator guys," Aubrey counters, "They keep at it in spite of the hurdles." iPal pauses, almost as if he's thinking. "Sounds like work...I'm just a toy remember?"

The night before, the large cargo module was loaded. In addition to the fifteen passengers and four crewmembers, 30,000 pounds of cargo will travel with Aubrey to orbit.

The elevator only holds four people at a time. It takes 45 minutes of waiting in line to take an elevator down to go up to orbit. Once on the bottom, the door opens to reveal a 20-foot wide room. A crewman greets him and sends him down the corridor through more hatches.

Instead of an airline seat, he is assigned a small cabin. It's a good thing; he has a 24-hour flight ahead of him. Entering the cabin, Aubrey notices there are no windows; in fact, he hasn't seen a window anywhere on the orbital airship. The scale of the vehicle hits him again. With the cabin in the middle of one of the "V" arms, he's at least 200 feet from the side of the vehicle. On the cabin wall is a big screen TV. Grabbing the remote, Aubrey turns it on. A menu appears. "Just like in the hotel," he thinks. The options are: External Views, Fore, Aft, Port, and Starboard, Station Schedule, Current Vehicle Position, and Movies.

There are seat belts on the chair. His instructor at his "introduction to space flight" class said that unless there was an emergency, he would never use them. Unlike the climb to the station, the trip to orbit will be very level.

A slight chime sounds and the captain's voice fills the cabin. "If you select the exterior camera starboard view, you may see something both common and rare. A large thunderstorm has formed about 100 miles to the south of us. If you watch carefully, you may see sprites or a blue flash. They are common events, yet it's extremely rare to see them with your own eyes. We hope you don't mind, we're adding an extra hour to the trip. We've been asked to take a sample from a Mesospheric ice cloud for the scientists."

After a few hours of climbing, Aubrey tries to see the point that they had entered space. On his handheld GPS he sees the official 100 kilometer number come and go, but on the screen there is no change.

"I think I feel lighter than I did this morning," Aubrey says out loud to his cabin mate. "I hate zero gee, but I love half-gee," says the man on the seat across from him. With a knock on the cabins sliding door, the steward is passing out electronic wrist bracelets that sailors use to prevent seasickness. He points out that under the seats, there is a small compartment. This is where the zero-gee supplies are stored: Velcro booties, small foam beanie hats, and space sick bags.

Settling in for the night, Aubrey stretches out on his bunk and dreams of Mars, while he floats to space.

Aubrey wakes up on the second day to something bumping his nose. He opens his eyes to discover his iPal floating in front of his face. "I'm in orbit," Aubrey whispers. iPal is, for once, speechless.

As Aubrey looks out the window at the blue planet below, a musing crosses his mind: "Sure is a long way from the river."

Chapter 15
Economics at the Edge of Space

If the ATO system is to be successful, its economics must be considered from the beginning. As the old expression goes, a ship by itself is a hole in the water in which to pour money, and spaceships are no exception.

It is not enough for the final system to make a profit. Each component—the first stage airship, the DSS, and the orbital airship—must generate revenue. Even the development systems need to make a financial contribution. This is the same need as a conventional transportation system.

When you ship a package, the courier service that picks it up makes a profit, the shipping company sailing across the Atlantic makes a profit, and the trucking company that makes the final delivery also must make a profit. Even the guy with the forklift needs to eat if the entire system has any chance at success. If any step of the process cannot carry its weight financially, then the system is in jeopardy. Additionally, the economics of the space industry are like a roller coaster. A solid market one day is an economic disaster the next. This has been the end of every non-governmental, long-term space development program. The solution is multiple streams of income. Each element of the system must have several independent market and income sources.

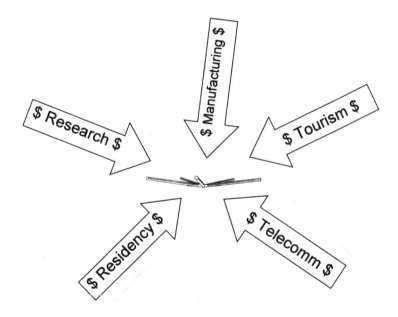

Figure 15-1: Multiple Streams of Income

Here are some of the potential markets for ATO components.

First Stage Airship

Broadband Communications

A first stage airship can act as the world's tallest telecommunications tower. Any tall tower use can be applied to a floating platform—television and radio relay, cell phone networks, and pager systems. The reach that a 25-mile height gives you is enormous. The line of sight from 140,000 feet is 500 miles. A typical 120-foot relay tower reaches less than 100 miles.

Obviously an airship is more complex and expensive than a steel tower. The cost trade-off compares single stations vs. multiple towers, land procurement vs. overflight clearance, purchased electrical power vs. station-provided power. The first tower-type application is likely to be broadband communications for rural cell phone networks and high value private networks, such as banking systems.

Remote Sensing

The first stage airship is an ideal platform for remote sensing. Satellite remote sensing gives only a momentary glimpse of a location. In those few seconds of sight, a remote sensing satellite can bring a wide array of powerful sensors to bear. However, what happens if you need to see for more than a few seconds? What if you need hours, or days? One of the early contracts for an Ascender airship was for the Air Force, under a project called "Constant Stare." The first stage airship can stay on station for weeks at a time.

Dark Sky Station

Port

Seaports are one of the quiet powerhouses of the world's economy. In 2003, two trillion dollars of shipping pass through United State ports alone. In spite of these huge cash flows, 30 percent of all ports operate at a loss. These ports are subsidized by government, both local and national, to keep the economic benefits of the high cash flow rate going. Like roads and bridges, ports are viewed as a critical infrastructure for a country's economy. When a solid customer base and good management combine, that port itself is a high-priced commodity. Recently, 50 percent of the container port in Shanghai, China sold for 1.6 billion dollars.

The Dark Sky Station's main purpose and income source is that of a port. However, like its seaport brethren, shipping alone may not bring in enough income to support costs. This will be especially true in the early stages of ATO operation when cargo is at a minimum.

DSS cargo will be of two types; cargo going from point to point across the planet, and cargo heading for space.

Tourism

Using a small island tourist destination as an example, it can be seen that tourism can make up a large part of a successful economy. However, the tourist destination island example also shows that a complete dependence on tourism does not provide sufficient income for growth or a high degree of quality of life.

Tourism generates dollars in two categories: the price of the trip itself, and the dollars that are spent while the tourist is at the destination. Dollars spent at the destination account for 80 percent of all tourism-related income. The tourism business model of the Dark Sky Station must be planned with this in mind. Current tourist trips to the International Space Station have no provisions for spending once there. There is no Starbucks on the ISS. The upcoming suborbital space tourism industry similarly lacks the ability to take tourist dollars beyond the price of the ticket. There is a common mistake in the developing space tourism industry, the belief that all passengers will want to do is to ogle out the window. This will attract paying customers for only a short while. Cruise ship companies learned this a long time ago. Even surrounded by the vast beauty of the ocean, passengers will opt to see a nightclub singer in the ship's lounge or play slot machines.

For the Dark Sky Station to be a real tourist destination, there must be a thriving economy at the station itself. Restaurants, gift shops, casinos, and unique activities must be part of the mix.

Manufacturing

The cost and availability of electrical power is becoming a huge issue when selecting a site for manufacturing. The solar power on the Dark Sky Station will provide cheap abundant power. At the beginning of the twentieth century, big aluminum producers moved their smelting factories to Niagara Falls. What drew them was the abundance of cheap hydroelectric power. In the mid-to later-half of the twentieth century, as power became more distributed, "locating to power" became unnecessary. Now as the price of energy continues to grow, locating to power is again becoming an important factor for industry.

Vacuum processing

The vacuum of near space is a major resource that can be exploited. At 140,000 feet, the air pressure is less than one percent of that at sea level. This is not "lab quality" vacuum; however, it is sufficient for many industrial applications. Many of the new composite materials require a vacuum.

The cost of a vacuum chamber increases sharply with size. Several cutting-edge aircraft, such as the Beachcraft Starship executive plane and the Boeing Dreamliner, use enormous vacuum-cured parts. In near space, extremely large parts can be exposed to vacuum. Composite structures such as entire airliner wings, bridge components, and stadium domes can be made as a single unit due to the unlimited vacuum available.

Residency

Like any small community, there will be personnel who pay for their apartments, staterooms, food, entertainment, and other amenities. This helps build an economic structure, although it is not an overall net gain for the station's economy. But life at the edge of space will appeal to more than the staff. A condominium at 140,000 feet could become the ultimate summer home.

Orbital Airship

Scientific Research

Scientific research is a multi-billion dollar industry. The orbital airship will have exclusive access to several fields of research. More and more discoveries are showing that the upper atmosphere plays a role in daily weather and climate change. The orbital airship will be the ideal tool for in situ research of the atmosphere above 140,000 feet.

Heavy Cargo Reentry

As other space industries develop, they will need a way to bring their product to market. With the pending retirement of the Space Shuttle, the only option is to stuff products into small reentry capsules along with the crew. This is fine if you make one space Beanie Baby a year; otherwise, it just won't do.

The orbital airship can carry back to Earth products manufactured in space. With the advent of mining and other heavy space industries, it is the only option on the horizon.

High Rack

High Racks are the balloon instrument decks used to test ATO systems at 100,000 feet. They serve as an example of support components, becoming part of the revenue stream. High Racks are already a major source of income for ATO development.

These simple stick, foam, and balloon vehicles are market builders and customer creators. Customers use High Racks as a low cost entry vehicle for near space-related business. As their businesses grow, so do their needs for bigger payloads. The hand-off from High Rack to DSS to the first stage airship is a natural transition.

Cargo

High Racks have carried trinkets, mementos, and even sunscreen to be tested under the high ultraviolet environment found at the top of the atmosphere. Learning the "ins and outs" of running near space cargo on this small scale provides invaluable lessons. When the time comes to carry tons instead of pounds, the system will be ready.

Advertising

Advertising dollars on High Racks has been another excellent source of revenue. On the High Rack, six cameras point at six tiny billboards. During flight, thousands of pictures are taken of company logos and messages. After the flight, the companies are sent incredible images of their logo with the curve of the Earth in the background. These are then used in magazine ads, websites, and even on the sides of their delivery trucks. There has been some controversy over space advertising. My answer is: if it's good enough for the football team, it's good enough for the science team.

The advertising and cargo customers covered the entire cost for the last five High Rack flights. The purpose of those missions—ATO development—got a free ride.

Figure 15-2: High Rack Ads

Mach Gliders

Intelligence

Even though a Mach glider is a research tool, it does have an end-product application, as an on-demand disposable reconnaissance platform. Being rocket-deployed, it can be onsite at any hot spot in the world in moments, take pictures, gather data, and then be disposed of. It's the throwaway spy plane. Commercial intelligence is also a prime customer for this technology. Private companies have become large consumers of intelligence. The first stage airship can provide in depth data within days, and Mach gliders can provide a quick peek in minutes.

Tandem Airships

Construction

Both real estate developers and upper atmosphere developers need their backhoes and cranes. The Tandem utility airship fills those roles. They will become the equivalent of the big yellow construction equipment in the sky. Tandems can be manufactured for new projects or could be leased out from the ATO development when not building Dark Sky Stations.

Training

There are dozens of companies designing and attempting to build airships that will fly at the edge of space. Most of these vehicles are flown by radio with the pilots on the ground. There is a problem. There are very few high altitude airship pilots. Before the prospective pilot takes the controls of the very expensive full size airship, he/she should crash a few cheap Tandem airships. Even without the crashing, Tandems can be effective low-cost training vehicles.

Transatmospheric Ascender

Point-to-Point Shipping

Even though the Transatmospheric airship cannot reach orbit, it does fly at Mach ten. This will make it the world's fastest cargo plane. The TA can deliver cargo anywhere in the world in as few as six hours. There is a major drawback for this use—the initial two-hour flight to Dark Sky Station and another two-hour descent from the station to the destination.

Suborbital Tourism

For the adventurous, the TA can provide short excursions into space well before the orbital airships are ready to fly. These 30-minute flights would start from the Dark Sky Station and could feature five minutes in zero gravity and a spectacular view for the passengers.

The income from the secondary market will allow time for the primary market, transporting of people and cargo to space, to become established.

JP

Chapter 16
The Challenges

Ideally, the entire ATO system will be built from off-the-shelf systems and materials. However, several new technologies will play an important role in ATO. Extremely exotic materials and technologies that exist only in theory (unobtainiun), should be avoided. But that prescription should not blind developers from taking advantage of cutting-edge technologies. Sometimes you need to move forward even if what you need is impossible. The invention of the jet engine is a story of facing the impossible.

While Sir Frank Whittle was inventing the first jet engine, the metals available weren't up to the job. Parts of his engines such as compressor blades just blew apart under the forces needed. Whittle continued anyway. He pressed on in spite of the fact that in his day the materials he needed just did not exist.

Sadly today most people in technology fields would have ridiculed Whittle. They would have shown on the back of an envelope that what he was trying to do was impossible and actively tried to have his work stopped. Unfortunately, this response is all too common. Only the wild and crazy ideas really change the world. For every "crazy" like Whittle or Wilbur and Orville Wright who succeed, there are thousands of those who don't. These are the unsung heroes of the future. Those that try and fail are the foot soldiers bringing in tomorrow.

The next time you hear someone criticizing a new idea because it needs a material or technology that is just out of reach, think about Sir Frank Whittle and remember that jet airliners really do exist.

Like the early jet engine, ATO requires a lot of hard issues to overcome. In the following pages we will examine these closer.

Size

These vehicles are huge. The first stage airship is hundreds of feet longer than the largest airships ever built. On the orbital airship each wing is over a mile long, and a Dark Sky Station can be over four miles across. Bridges and dams are the only structures built by humankind that are of this size, but they don't float or fly.

The primary difficulty in building structures of this size is actually psychological rather than one of engineering. One of my favorite things to do when showing the small Dark Sky Station test vehicle is to ask someone to lift one of the arm segments. The arm segments are stored standing up on end, so the three-foot wide structure towers twelve feet over the victim. The person will grab vigorously on the framework and give a hefty lift. The result is always the same. They nearly throw it through the roof. They expect it to be much heavier. Even engineers who have been told the weight and know the materials used have the same reaction. As gigantic lightweight structures become more common the idea of the vehicles miles long will be seem less startling.

Even the early development process is impacted by size. The hanger that was used to build the Ascender 175 cost $35,000 a month to rent. The Ascender 200 will need an even bigger building. When the Ascender 900 is ready to be built, it will need a truly enormous custom building.

Figure 16-1: Arm of a "baby" first stage airship

Drag Reduction

Managing the plasma flow around the airship at hypersonic velocities is critical to drag reduction. The lower the drag, the less thrust will be required. If conventional drag projections using supersonic flow models are used, the projected drag is too great to overcome. The airship would never reach orbit. The low drag techniques must be verified. Gathering the data from the high altitude Mach Gliders will be critical to creating the correct aerodynamic model.

Assembly at Altitude

The need to assemble float structures while they're floating is challenging enough. Added to that is the challenge of assembling them in the harsh environment of near space. Entirely new construction techniques must be developed. Skill will be needed that we are not even aware of yet. Like many aspects of ATO, the hardest part to overcome is that barrier in our minds of those things that have never been done before.

Difficult to Simulate

Most aircraft today are completely simulated on a computer, not just before the first test flight, but often before the first part is made. This is possible due to the vast amounts of data and mathematical models for conventional aircraft flying in the lower atmosphere. That level of knowledge simply does not exist for large-scale non-conical hypersonic vehicles. In addition, there are gaps in our understanding of the upper atmosphere. Programs are in place to investigate the known gaps. The challenge is responding to the true unknown, factors that will take us completely by surprise and those established facts that turn out to be false.

How do you model the unknown? You gather data. For ATO that means throwing inflatable stuff into the upper atmosphere and driving it fast.

Funding a new and different concept

In spite of the promise of upper atmospheric infrastructure, it will be a difficult project for which to raise adequate capital. Keepers of the purse strings tend to be a conservative bunch. Traditional funding sources are a bad fit for programs like ATO. ATO's long development eliminates venture capital and stock sales. The high risk leaves out banks and institutional investors.

One way of getting funding is to go directly to the market—not with ATO, but with the parts. Along the way, components, systems, and entire vehicles will be developed that have their own commercial application. The perfect example of this is the dual-balloon airship. This development tool for ATO has applications for the remote sensing and telecommunications industries.

The greatest challenge is to not be lead astray, to both stay on track with ATO while taking advantage of the commercial opportunities.

The Environment

Large-scale electrical phenomena such as sprites and gnomes are also an unknown risk. They could be merely a beautiful part of nature to see, or they could be devastating to anything in their path. Balloon and unmanned airships flying directly into these events will be required to answer this question.

Regulatory

There is currently a complete ban on flying unmanned aircraft in the US. The only exceptions are those vehicles within Department of Defense programs and small hobby planes. The big concern is potential collisions with manned aircraft, i.e. the drone vs. the 747. The Federal Aviation Administration has laid out long-term plans for coming up with new airspace rules to accommodate remote-controlled vehicles. However, it looks like new rules and clearances are many years away. No one knows what the final regulatory arena will look like. It could be positive, or it could be hostile. Using the past as a guide, the new rules will be at first awkward and difficult to accommodate, then slowly improve as they are used.

ATO vehicles have pilots on board; however, the early development vehicles are flown remotely. Regulatory issues already consume huge resources. The government adding a whole new book of regulations is only going to make it harder. Options include conducting the tests outside of the United States or putting people on board much sooner. We hope that neither option should be necessary.

What to do

You do a step-by-step approach. Work out each problem as it appears. I have found that "impossible" means no one has done it before. Completely impossible means you need to find a way around.

I have seen countless examples in my lifetime of "impossible" things being achieved. They were not just idly said to be impossible, they were mathematically proven to be impossible by experts who really did know their field. Look closely at something that was done that was proven to be impossible. Very often the key reason it was considered impossible still exists. The problem was overcome in an unexpected way.

When engineers first started designing spacecraft reentering the atmosphere, they had a big problem. The temperature the spacecraft would reach would melt or burn any known material. This was the hard fact. The engineers had two choices; they could spend the next few decades trying to develop a super metal that doesn't melt, or they could make an elegant turn, and fly around the problem. We have space travel today because they did the latter. They discovered that if the spacecraft had a blunt shape, a shock wave would form in front of it. Most of the heating would occur at the shockwave and not on the spacecraft. The reentry capsule also employs another trick—if you can't stop the material from melting, let it! As the surface of the heat shield melts, it is blown off the spacecraft, taking the heat with it. The principle that made reentry impossible helps make it work.

Were the naysayers wrong? No, their numbers were correct. Were the original facts wrong? No, those reentry temperatures are just as valid today. Was reentry impossible? Ask a Space Shuttle pilot.

JP

Chapter 17
Working the Program

At 3:00 am the alarm goes off after only two hours of sleep. I never sleep well before a mission. At 3:30am I perform the opening act of the performance that is a typical mission—knocking on motel doors and rousting the team. Instead of grumbling I'm greeted with smiles and even a few awake people. It's a good sign for the day ahead.

Thirty minutes later the team is loitering in the gravel parking lot of Bruno's, the former horse stables, now resort motel of the desert here in Gerlach, Nevada, on the edge of the Black Rock Desert. It's dark, but still warm from the day before. It's going to be hot on the lakebed today. After a head count, we climb in the cars and caravan over to the launch site.

While rolling to the desert, the mission plays through my mind. "The guys look ready. Should we push back antenna setup until after balloon bag setup? Can we gain five minutes out of moving the tank by leaving in the truck later?" "No, no, no, just go with the plan." On and on I play the mission in my mind until it's time to pull off the highway and onto the lakebed.

We're flying two missions today, Away 32 and Away 33. Each is a seventeen-pound collection of systems, batteries, carbon fiber, and foam. Both will be carried to over 90,000 feet on their own balloon. Both of these flights are part of ATO development. There are lots of small steps toward ATO. The Away missions are the tools that accomplished many of those steps.

Away 32 will test telemetry system improvements. The upgraded communication systems are critical to all aspects of the ATO system. The same is true for the precision fill system and the high wind launch bags. The new balloon adapter is being tested for the Tandem airship. The Tandem isn't directly an ATO vehicle, but will be a workhorse support vehicle.

Away 32 is also a financial mission. All around the vehicle are miniature billboards. Six digital cameras on carbon fiber booms point back at the seventeen ads. The ads cover the costs of both missions. Integrating commerce into development flights is the real key to making ATO possible. This process can be a bit tricky. In order to do it right on the big DSS missions, we need to learn the process on the small missions first. In this way, the commercial payload is also a critical research tool.

Away 33 is carrying a load of ping pong balls. These experiments are called PongSats. Students take a ping pong ball, cut it in half and put an experiment inside. We get PongSats mailed to us from all over the world. After the flight, we mail them back, along with data and video from the mission. Experiments range from marshmallows to computers with arrays of sensors.

The PongSat program is not only a learning tool for the students, but for us as well. We're learning to deal with huge numbers of customers and getting a taste of bulk cargo. When you are managing hundreds of individuals participating in your mission, you either get good at it fast or you lose your mind. I'm not entirely sure which option we've taken. We've flown over 3000 PongSats. When I step on the moon I will have PongSats in my pocket.

Economics have driven us to fly multiple missions on each trip to the desert. If you reuse and recycle your vehicles, the majority of the mission costs are in logistics. There's the cost of rental trucks and trailers for hauling all the gear to the desert. There are fuel costs and everything involved with supporting a team of people in the desert for a few days. All of these costs are the same whether you fly one or three vehicles. To pull it off, several development tracks must be run in parallel. Your flight team needs to be good. The morning launch window doesn't get any bigger because you've doubled the number of things you're doing.

There is a less tangible, but no less real reason for doing these flights—to learn to work in the environment of the upper atmosphere.

It's still completely black as we roll off the highway onto the dry lakebed. We pull just far enough in for all the vehicles to fit in a single file line and then wait. As soon as the last few cars catch up, we begin the crawl. If you go more than seven mph you pick up dust. It's fine in the front, annoying in the middle, and a snowstorm like whiteout conditions in the rear. At exactly one mile on the odometer we stop again. It's time for a walk.

Cars have been known to sink up to their doors in the mud of the lakebed. The problem is you can't see the mud. It can hide half a foot below a deceptive hard cracked surface. You can tell if you get out and walk it, although it is harder to tell in the dark. I take a flashlight and walk 100 yards to the side. I reach down and pull up a piece of the cracked ground. There's more water on Mars. It's time to roll in.

The first vehicle that pulls in is the mission control van with the equipment trailer. We position is so the open side door will face the lift-off area. This puts all the antennas on the other side, keeping cables away from tripping feet. The rest of the operation forms around mission control.

As soon as we get parked we have the team meeting. There are sixteen of us this morning—eleven team members and five observers. No floodlights are up yet, so the team huddles in the glare of car headlights. Everyone already knows their tasks, yet when it's cold, dark, and 4:30 in the morning, it doesn't hurt to have a quick review. Like the huddle before a football play, the meeting is a chance to get everyone focused and in sync with each other. The words said are really secondary. I hand out the mission sheets, and we do a quick rundown of each flight and team task breakdowns.

"The theme of this morning is "pacing", I begin, "We have two complex vehicles to prep and launch in the next few hours. I want everyone to focus on pacing."

"We need to keep moving to get everything in the air before the forecasted 30 knot surface winds come up, but don't want anyone rushing. Think about your pacing. I'll bounce between the teams and bump you up or slow you down so you're in sync."

Everyone now has their mission sheets. These have all the basic information the team will need— flight weight, how much helium to put in each balloon, lift off time, balloon rigging diagrams, frequency lists, etc. A week earlier everyone was already sent the same info. The hardcopy acts as a "cheat sheet" that the team keeps in its back pocket.

The first step is deployment. We review who will be getting what out of the van, trucks, and trailer. Then comes the first split. Part of the team will begin setting up mission control, and the others will set up the antenna farm. The team gets their tasks and they're off. "Next meeting here in 45 minutes, lets do this!"

The team heads into the dark, and the mission begins. There's a low rumble and the generator comes to life. Within minutes, halogen worklights shine on the desert. Everyone is moving with a purpose. Awnings and tarps go up in front of the mission control van.

After unpacking, the next task is to set up the antenna farm. There are seven antennas to set up. Four of them are huge sixteen-foot long yagis. A yagi antenna looks a lot like a rooftop TV antenna. The big yagis are mounted on swivel stands that allow them to be pointed at the balloons during the flight. Both the stands and the antenna now need to be reassembled and cables run to mission control.

The antenna team has divided into two, one for each of the big stands. The first team is really going to town. The second team has a problem. The main pole of the antenna stand is missing. We had set up the antenna in our parking lot only a week before to verify that all the parts were there. How could the main mast be missing? Two seconds for a heavy sigh and then it's on to "what we are we going to do?" These antennas are mission critical. They are too big to just hold up and they cannot be just leaned against the mission control van with all its radio interference. The antenna team has lots of ideas and a very determined attitude. It's time to walk away from this issue and let them get to it. Every mission will have its issues and problems. We actually plan for them. In the mission timeline, there are lines like "problem one, fifteen minutes." You never can predict what will go wrong, but if you assume that things will, you can be ready.

As I'm walking over to touch base with some of the observers, the lakebed gets quiet. Problem two has just shown up. The generator has conked out. Bob walks over. We discuss options—battery to jump the generator with the backup deep cell battery, cycle cars, put up the solar panel when the sun gets up, run on minimum systems, and others.

I leave Bob to it and grab Ed. He's leading the Antenna One team but will change roles soon to become the Balloon Captain. We make a few changes to the upcoming events, and then he's back on the antenna.

Kevin and Paul have set both Away 32 and Away 33 down under the awning in front of mission control. Kevin starts laying everything he'll need to get the vehicles ready for flight. The prep table is set with quick ties, tape, voltage meters, diagonal cutters, and scissors.

Bob is the flight director for both missions and he's buried in the van getting all the systems up. Our mission control van is a former TV news van. It's lined with racks of radio gear, monitors, tape recorders, and computers. It has its own generator and power systems. It's our own self contained "Houston." A few weeks ago we completed a major overhaul of the interior. This is the first field test of the new layout.

Time to break out the checklist. During a mission, I carry around a mini file box. It holds all the mission paper work, clearances, weather forecasts, every possible mission-related phone numbers, mission sheets, and the checklists. I grab Kevin and Bob and hand them each a copy of the launch checklist for Away 32. Kevin has the "checklist" for this mission. His checklist is the real deal. Mine and Bobs are for reference only. Kevin writes MASTER on the top of his and asks me for the time. "6:10am", I say as I read off my watch. Kevin writes the time in the "start time" box on the checklist, and Bob starts the first mission clock. The launch process has begun.

Even though we are a volunteer group, we get very formal at moments like this. The formality provides structure and clarity. When "mission chaos" sets in all those little formal moments really pay off.

Two folks are assigned to continue fixing the antenna problem, and I gather the balloon team. I cancel the general meeting and formally hand over leadership to the balloon fill captain. Meetings are really a breaking point for changing tasks and restructuring teams. They last no more than ten minutes, but they are vital. It's a chance to reevaluate, catch problems and get the feel of the team. During the meeting I notice that the generator is humming away. I can't remember when it started but I mentally cross it off of the list of issues.

The first task for the balloon fill teams is setting up the launch bags. They lay there, looking like overgrown sleeping bags. The launch bags act like airplane hangers for the balloons. Each balloon is filled with helium inside the bag. When it's time for launch, a long Velcro "zipper" is pulled, opening the entire top of the bag. The balloon pops out, and up it goes.

The balloon team has two launch bags to set up. The first to be laid out are the large tarps, then the bags themselves. The team has spent six months sewing these bags together. There's no need to tell them to handle the bags carefully. Lots of stakes, 176 in all hold the launch bags down to the desert floor. Each bag has 216 pockets. Jill and K'John are sitting on the bags in their socks filling the pockets with bean bags. The bean bags add weight across the top of the balloon. For every launch, the pattern and number of bean bags are changed to match the launch configuration of the balloon.

With the pockets stuffed, the sock wearing crew now attaches the tear panel. The tear panel is a strip of nylon eighteen inches wide and 30 feet long. It has Velcro running on both of its long sides. Then Jill and K'John careful match the Velcro on the tear panel to the matching Velcro on the launch bag. It needs to be flat, straight, and uniform for a good launch.

Ed crawls inside the launch bag and lays out the balloon. Uninflated, it is a twelve-foot long mass of rubber. It's pulled into line directly under the tear panel.

The horse trainers have their Horse Whisperers; we have Paul, Balloon Whisperer. Paul attaches the carbon mount to the balloon and gives it all a once over. A little bit of tape here, a quick tie there, now he and the balloon are ready for helium.

One of the purposes for this mission is to perform a field test of helium metering. We need to know precisely how much helium we're putting in the balloon. Traditionally for small balloons this is done with a spring scale. As the helium flows in, the balloon lifts up on the scale. The lift is directly measured and a simple calculation determines the amount of helium. This technique has two problems—one, any wind throws the scale off. The top of the balloon acts like an airfoil on a wing. Even a one-knot breeze can cause the scale to be 20 percent off. The second problem comes from the use of the launch bag. You can't float the balloon to measure the lift. To get around all this, we are trying a flow meter to measure the helium directly as it goes into the balloon. The meter is the type used for natural gas in industrial buildings. Standard helium meters simply do not have the flow rate we need for balloon filling. We tested it in the shop so we can compensate for the difference in readings between natural gas and helium. Unfortunately, the readout is a bit of a challenge. For each half a cubic foot of gas the indicator needle makes one revolution. The next needle counts in hundreds of cubic feet. The second needle is too coarse. We need to make a check on the sheet for each half–cubic–foot needle turn. If you look away for a moment, the fill will be off. Shelly has taken on this job. She is staring at the meter so intently I think it's going burst into flames. It goes well, but we definitely need a different meter.

I'm looking at the balloon and it doesn't look quite right. I quiz Karl about the fill. He looks at his checklist, smiles and says, "I'm pretty sure we're right on." That's Karl–speak for "I absolutely know for sure it's correct." Under-filled launches on windy days are dangerous. I make the call and ask Karl to add another fifty cubic feet of helium.

Checking back at the antenna farm, I see a weird thing of beauty. From the parts of the old stand and spare bits from the van, the team has invented an antenna stand that just may be better than the original.

Back at mission control the vehicle is coming to life. One by one, systems are turned on. First the battery is tested for the proper voltage. It then gets plugged in and the system is powered up. Each system beeps when it's first turned on to say it's "OK." After the beep check, mission control verifies receipt of the transmission from the system. The power switch gets taped into the "on" position, and then the unit is sealed. For Away 32, there are five pages of items to turn on and check. Every mission will have a systems glitch, a GPS that won't see its satellites, or a transmitter will reset the computer of another system. No matter how much you test, there is always a new twist in the field. This time the vehicle powers up without a hitch, spooky.

One of our traditions is to have all the members of the team sign the fin of the vehicle before launch. We cannot stop the flow for everyone to sign at once, the marker gets passed around, and one by one the team members dash over to the vehicles and sign a fin.

Kevin is calling out the last of the checklist items—foreign object removal, balloon team go, notify balloon team of impending vehicle attachment. In the balloon world this is when the vehicle is "hot." Ground camera go, mission control go, sky clear of traffic, I look Tracy in the eye, he's "go."

I get a checklist complete from Kevin and call out "go" for flight. The wind has picked up to ten knots. Tracy and I lift up the vehicle and get ready to do the launch dance.

After the final go, the actual launch command is given by the captain of the balloon team. With everyone ready, Ed stands holding the anemometer, (wind speed meter). He's timing gusts. He calls "go" and I hear the tearing of Velcro over the wind. The launch bag turns from blue to pure white. The balloon has popped out. The wind has caught it and is pulling it to the side instead of over the vehicle. Tracy and I twist the vehicle around and take two steps toward the balloon. The vehicle is pulled from our hands and it's up and away.

We all watch the flight for a few seconds and then it's handshakes all around. Within three minutes, the teams are starting it all over again. The second balloon is being positioned inside the launch bag, and the checklist for Away 33 is started.

Poking my head inside the mission control van, I ask Bob how fast is Away 32 going up. It's climbing at 1,300 feet per minute. Way too fast—the projected climb rate was much lower. There is too much helium in the balloon. A moment with the calculator shows it to be about fifty cubic feet too much. Karl was right. Instruments beat eyeballs every time.

The second fill has already begun. The team has opted to reuse the first launch bag instead of moving the second bag. This means they don't need to move the fill gear. The balloon team normally is ready to fly before the systems team. There is always a glitch that needs to be tracked down on one of the vehicles. The two launch bags allow for the first balloon to be left unattended while the team fills the second balloon. This time both the balloon and the vehicle were ready together.

About halfway through filling the balloon, one of the stakes holding the bag down pulls out and strikes the balloon. The balloon doesn't burst. A quick inspection shows no damage. Two stakes replace the one, and the fill continues.

Away 33 is carrying 320 ping pong balls, each filled with a student experiment. These are part of the PongSats program. PongSat is short for ping pong ball satellite. Weeks before a mission, students from all over the world start mailing us their handiwork. We carry them aloft, then mail them back along with mission data and the on board video. Thousands of kids are running their own space programs through PongSat. We don't charge for the ride. If any readers want to fly to the edge of space, go to www.pongsat.com.

Most of the PongSats are passive; however, twelve have on board computers and sensors. The active PongSats need to be switched on and their power up times noted. Four of them are on the upper deck with sensors pointed at the sky. The others are mounted in two rows on the lower deck. As Kevin called out the PongSats ID number, I switch it on and call out the time. The hundreds of passive PongSats are in an upside down pyramid shaped bag suspended below the upper deck.

We're ahead of the game so we hazard a moment for Natalie to take some team pictures next to Away 33. We're carrying a quarter pound of Vista Clara Coffee on each vehicle. Coffee bags are handed out to the team and the sponsor pictures are taken. The first picture has all the team members holding the coffee bags with bright smiles. Happy sponsors mean more flights. On the second picture we tell everyone to be themselves. This means hand gestures behind people's heads, gnawing on coffee bags, and other assorted unpublishable abuses. Heavy sigh... this crew is the best, but you just can't take them anywhere.

Vic is from the aerospace giant Lockheed Martin. He came up with his coworker Chris to observe the flights. We hope they will participate on an airship flight we have later in the year. After the picture taking Vic, asks if it's alright if he puts a Lockheed Skunkworks sticker on the side of Away 33. The Skunkworks is this near mythical, legendary group that has created some of the most amazing aircraft ever flown. It was a great personal moment for me that they want to put their sticker on my vehicle. I gulp, smile and try my best to be nonchalant, "Sure, anywhere you like".

We're down to the last few items on the checklist. It's time to move Away 33 out to the balloon.

In the movie *Apollo 13*, one of the most exciting moments is when the mission controllers call out "Go!" one after the other. There's nothing like being there and calling out one of those "Go!" yourself.

Once again at Ed's call of "Go!", Paul makes his run. Off goes the tear panel and the balloon is sent sailing upward. A cross wind catches the balloon pushing it to the side of us instead of above us. Tracy and I try to take a quick step to get under the balloon, but the vehicle pulls away from us before we get a chance. Away 33 pulls out of our hands and takes a big swing toward the ground. It misses by a good eight feet and continues to rock for a few minutes before it settles out.

The launch bags can handle very high winds when they are firmly stapled to the ground. They are pretty fragile when loose. A few missions back both bags were torn while putting them away when gusts came up. The forecast for the afternoon called for winds up to 40 knots. I turn around to see the balloon team already has the launch bags stuffed back into their boxes. The vehicle has been in the air for less than a minute. The team was really "in the zone" today.

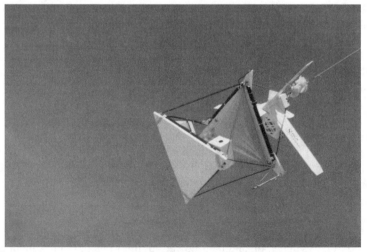

Figure 17-1 Away 33 in flight

I run to the mission control van, where everyone is asking Bob for altitude and system status. I ask everyone to give him some room to do his job. After all, he now has two vehicles in the air to manage. Actually, I just want to ask him my question, "what's the climb rate?" Bob waits patiently while I wait impatiently for a pair of position transmissions. "850 feet per minute", he announces. Exactly as planned, and I breathe a sigh of relief. There is nothing I need to do for the next fifteen minutes. I wander away and hide in the shade of my truck for a moment's getaway. Two minutes later I hear, "Have you seen JP?" Break over.

Away 33 carries a live video transmitter. Everyone takes turns looking at the monitor in mission control. The four camera heads give quite a view. About every ten minutes, the image on the screen gets snowy and a few bad characters show up in the data. It's time to re-point the antennas. We employ a very sophisticated method of antenna alignment—while Kevin shouts "a little to the left," Tracy obliges by steering the sixteen footers. "And now a little up," the call goes out again, "No, no too far...wait! Perfect!" Then all the antennas will be aligned with the first. This will go on every ten minutes until both vehicles have landed. There is an automated antenna tracking system in the works, but it's a low priority and keeps getting delayed. I think the crew just likes messing with the antennas. Some of them pride themselves on having "the touch."

As Away 33 passes 50,000 feet, something is wrong. It's slowing down. The climb rate is down to 500 feet per minute. There's really only one way that can be happening; we're losing helium. Maybe the stake-hit on the balloon did some damage we didn't see.

Burst! The balloon lifting Away 32 has popped and the vehicle is on its way down. Bob sends up command to release the remains of the balloon. The balloon reached 94,000 feet. This was lower than planned, but matches exactly what is expected from the overfill of helium. Away 32 accelerates rapidly downward in the thin air at high altitude. It passes Mach one before starting to slow down.

The day before the flight, we run a computer simulation that shows the flight path and the landing spot. The winds can change in the 24 hours before launch, but it gives us a starting point. We then take a hard look at the regional weather map and make our best guess where everything will come down.

On a mission the previous November, the vehicle flew 120 miles downrange, a significant distance when traveling across the Sierra Mountains. This time the prediction put the landing site just ten miles to the North of the launch site.

As the Away 32 descends it is only a few miles away. The team is getting excited, a chance to see the landing! The normally calm and disciplined crew members now dash to their cars and zoom across the lakebed. Only a few of us remain with mission control, trying to get a handle on the situation. We realize immediately that no one driving down the lakebed has the latest coordinates. Worse, the vehicle is moving west at a good clip and will land nowhere near the lakebed. About twenty minutes later, everyone else realizes it also and the cars come rolling back in.

Time for a team meeting. It's hard to jump down on a group of people that have just performed two perfect launches; however, chaos in the desert can undo all the morning's work if unchecked. The chastised team drops back into "mission mode." Natalie and Ed, the two recovery drivers, sit down over the map and plan the recovery. By now we have a good projected landing site for Away 33. The plan is for each recovery team to get as close to the vehicles as they can by road, then hike in the remaining distance. "Turn back time" is set so that both teams are out of the field before dark. We arrange a backup radio rendezvous point. If a team is not back by 4:00 p.m., an additional team will position themselves at a high point between the two landing site and call out on the radio. If nothing is heard by 5:00 p.m., then the search for the recovery team will begin. This allows the recovery teams a fall-back position. If the package is in a hard-to-reach spot, they can get into position and wait for additional help. If they get into trouble, they don't need to go for help— help will come to them.

While the entire recovery hubbub is going on, Bob is still in mission control managing Away 33.

Its climb has slowed to just 200 feet per minute. The balloon is six miles to the northwest at 90,000 feet. The video system is not holding up well. One of the four cameras is down. The other three are so snowy that little can be made out. This was expected of the vehicle when more than forty miles downrange, but not when it's nearly overhead. The one image that can be made out is the straight view of the balloon. The big white ball is hard to miss. We estimate that with the original helium load, Away 33 would have reached 102,000 feet. We're loosing helium faster and faster. We realize we're not getting much higher today.

At 22,000 feet, Away 32's descent rate slowed dramatically. The parachute has deployed. During the flight, the parachute sits in an open-topped container that has holes in the bottom. It's about the size and shape of a shoebox. In fact, we could probably use a shoebox and save some construction time. When the vehicle is on the way back to Earth, fins and the way it's balanced keep the base pointed down. Air blows through the holes under the parachute. At high altitude, the vehicle will fall faster than Mach one. However, there still isn't enough air to push the parachute out of the box. If you have a mechanism to artificially deploy the parachute, it would just flop around, getting tangled in equipment. When the vehicle drops low enough that the air is thick enough to support the parachute, out it comes on its own. This way the parachute only deploys when it can do so with a solid amount of air to keep it taut above the vehicle.

Away 32 is down. The last solid fix was at 1000 feet above the ground, sixteen miles away. The first team heads out. The second recovery team holds off. Away 33 is on the way down and fifteen minutes from touchdown. The exact landing coordinates override the desire for a fifteen-minute head start. Away 33 transmits one last location 1000 feet about the ground. There is a row of hills between us and the landing site, blocking all communications. If the wind continues blowing at the same speed all the way to the ground, touchdown should be 400 feet west of the last fix. With the landing position for Away 33 in hand, the second recovery team rolls.

With the vehicles down and the recovery teams rolling, it's time to pack up. It's over 105 degrees, the equipment is dusty, and the team is tired. I think the only thing keeping the tear-down crew going is the promise of air-conditioning back in town when they're done. After an hour of disassembly and packing, we walk the desert inspecting the site. The Black Rock Desert is a sensitive wilderness area. We follow the "leave no trace" policy and pick up every piece of tape and quick-tie we bring. The team has commandeered a big table at Bruno's restaurant, and with laptops and ice teas all around, we take the first look at the flight data. Bruno's world-famous raviolis are on their way.

The first recovery team rolls in at 2:30pm. Away 33 is in the back of the truck looking a little crunched. The frame is designed to break and absorb the landing loads. A quick check finds all the systems turned off and in excellent condition.

Away 32 landed ten miles further out and in a harder-to-reach area. The recovery team has found the vehicle, but they're exhausted. These are the same folks who were up at 3:30 am and launched the balloons. Now the 100-degree heat has gotten to them. At 4:00pm they call with their status to the team waiting at the holding point. You can hear their smiles over the radio when we tell them that we have a cold sodas and a relief crew waiting.

We follow them back out to the landing site. The cars can get within a half-mile. The rest will be on foot. The first thing we do before even touching the vehicle is take pictures of the site. Next we power off all the systems. In its broken, but connected state, Away 32 is awkward to carry. It takes three of us to hold it while walking and stumbling over the bushes and dunes. Tracy starts up a chorus of Marine marching songs. I can sing or I can march, but I can't do both. K'John takes my place and we head back to the cars.

It's nearly 8:00 p.m. by the time we get back to Bruno's. The team was going to spend the night in Reno before heading back to Sacramento, but no one is up for the two-hour drive. We get our rooms back with a promise that the air conditioning will be working tonight. The banter around dinner is always the same after a mission..."Can you believe we launched in that wind!" "What is this bump in the data?" "Next time I'll…..".

After dinner, we pull all the cameras off both vehicles. That night in the motel room I download all the ad images from the digital cameras onto my laptop. The images look good. We financially live to fly another day. I'm now both wired and completely exhausted. I don't really fall asleep, I pass out. On Sunday the crew makes the six-hour trek home. The team is slowly dispersing along the way. Some are catching flights and others are off returning rental vehicles and equipment. By the time we make it back to the shop, just a few diehards unload the gear and finally put the mission to bed.

Monday hits with the realization that it's not over yet. Several schools are starting summer break on Friday and need their students' PongSats back by Wednesday. The answering machine is filled with ad customers wanting to know how their pictures came out.

First to the PongSats; they need sorting, certificates, pictures, datasheets, and video. My office looks like a ping-pong ball factory as PongSats are sorted into their groups. Scenes from the three ground and one onboard camcorders are edited with the transmitted video to make an eight-minute documentary. The next step is a marathon and requires preparation. I get a full pot of coffee, five ink pens and the "Sons of the San Joaquin" playing on the stereo, and I'm ready to sign 320 PongSat participant certificates. I'm a habitual writer to my congressperson and the President. After many years of replies with stamped signatures, one letter came in with a real handwritten signature on it. It was from President Bill Clinton. I don't know if I effected a policy change, but I do know that the letter is more meaningful because of the real signature. After that I vowed to never use a signature stamp. The price is writer's cramp at around signature number 280.

Economic viability must start during development. If the bill don't get paid, nothing flies. Each of the six cameras on Away 32 took four hundred pictures of the billboards around the vehicle. After trimming off the twenty landing site images of the same bush and a few at the beginning with unfortunate images of people standing next to the cameras, the pictures are put on to CDs and sent to the customers.

About 40 boxes head out, carrying small slices of the edge of space to all over the world. Now the mission is over...well, except for the analysis.

What's this got to do with space? It was just a couple of balloon flights….

This is how ATO will get built, one mission at a time. Developing ATO takes a lot of new systems, new technology, and whole new ways of doing things. We literally have a giant list. Each flight, another stack of things gets lined off. Sometimes the results make us add more things to the list. It's working the problem, step-by-step, mission-by-mission.

Some of the things we do on a mission directly apply to ATO. The Away 28 balloon deployment system is the very mechanism that will be used in the large Dark Sky Stations. The telemetry system tested on Away 32 will be used on all ATO craft, from ground to orbit. The helium metering we used on both Away 32 and 33 will be used throughout the system.

Some of the things are for tools we need to develop ATO. Several systems and structural elements on Away 32 are from our new airship. This airship is not directly part of ATO; however, it is a critical tool that we need to move forward. This mission gave the parts a critical shakedown. Sometimes you need to make the hammers and saws before you can build the house.

When ATO becomes a reality, tens of thousands of science and engineering-trained people will be needed to take advantage of the new reach to space. There is already a huge shortage of scientist and engineering students. That's why there are PongSats on every mission we fly.

The edge of space is our ladder to the stars. We need the experience of "being there". The more time we spend in the upper atmosphere the more we learn, the more we know. The knowledge and experience gained has such great value, it cannot be measured. Before we fly though the upper atmosphere at Mach 24, we need to be "old hands" in that ocean, the place needs to be our home stomping ground.

There are no impossible things, only to-do lists that are so long people are afraid to work them. We're tackling ours line-by-line. This is how ATO will get done; this is the countdown to launch. It's the long road that only the stubborn and persistent travel.

Figure 17-2 The Team

Appendix:
The Missions So Far

JP Aerospace is the developer of ATO. Here are the development missions, what we've thrown up into the sky so far. We believe in the "build a little, fly a little" method of development. More is learned from a rocket in the air than a rocket on paper. The process tends to consist of a lot of small incremental steps. The result has been the accumulation of a great deal of experience. What follows here is a summary of those experiences. Over the years, we've had our share of great success and great failures, and we've benefited from them all.

The test flight is just the tip of the iceberg of the process of developing space systems. They do serve as a chronicle of events. These are our efforts in pushing back the frontier of independent space exploration.

Over 89 development missions have been flown. *These are the voyages....*

First Line-Suspended Rocket Test Launch Series   **12/26/93**

Our first step was to use balloons to carry a rocket up high before launch. This combination of rocket and balloon is called a "rockoon." In rockoons, the first problem you encounter is "how do you launch a rocket hanging from a string?" This is when we set out to find the answer. A series of three launches were executed. A small rocket was used to study methods of launching a rocket from a platform suspended from a line. All subsequent launches have used the fundamental dynamics learned in these flights.

First Free Flight Rockoon Launch   **01/21/95**

Two dozen party balloons and that same small rocket and the rockoon was reborn. In this test, a small rocket was launched from a free-flying launch platform carried aloft by balloons. The balloons carried the platform to an altitude of 300 feet. A small computer on the platform controlled the launch of the rocket, and afterwards it released the balloons. The platform landed by parachute.

ATO is fundamentally an oversized rockoon. This twelve-inch rocket taught us things that apply to the orbital Ascender.

Medium Rocket Line-Suspended Launch   **01/21/95**

In this test launch, a four-foot rocket was launched from a platform. The platform was an enclosed rail made of paper and foam. The platform was suspended from scaffolding. The rocket launched cleanly from its foam rails and flew to 3,000 feet.

Medium Rocket Rockoon Launch   **02/25/95**

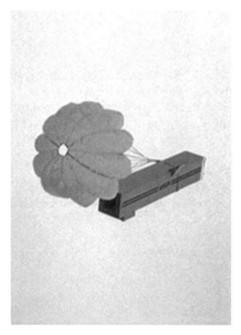

In this test launch, a four-foot rocket was launched from a free-flying launch box. Two trains of red balloons carried aloft the launch box. When the rocket was launched, its motor "chuffed," meaning it sputtered on ignition. This melted the foam around the rocket, preventing it from leaving the platform. The rocket motor kept burning and the launch box caught on fire. The burning rocket and platform landed by parachute and was quickly put out by the launch team.

Eighteen Test Stand Motor Firings   **03/14/95**

The "chuffing" rocket motor problem had to be solved. We developed several new rocket motor ignition methods. We purchased a large number of rocket motors and proceeded to fire them all on a test stand. We finally found a reliable motor ignition method that would not "chuff."

Medium Rocket Line-Suspended Launch w/Redesigned Platform   **04/08/95**

This was the second in this launch test series. A five-foot rocket was launched from a platform that was suspended from scaffolding. To protect the platform from hot gases, we designed exhaust deflection panels at the top of the platform and around the base of the rocket motor. The rest of the launch platform was redesigned so it could not catch on fire regardless of what the rocket motor did.

Medium Rocket Rockoon Launch Attempt   **05/21/95**

The mission was aborted shortly after liftoff due to high winds. This was the beginning of our struggles with the wind. While fighting with the wind-whipped balloon, we held a large sheet over it. We didn't realize we were glimpsing at one of the answers to the wind problem. It would be another eleven years before we will take advantage of the idea.

Medium Rocket Rockoon Launch   **05/28/95**

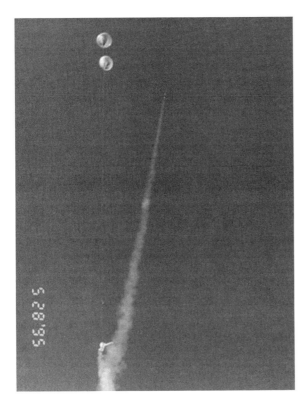

The launch platform was carried aloft by large weather balloons. This was our first use of weather balloons. The balloon carried the launch box to the 500-foot launch altitude. This was a beautiful rockoon launch, no wind, and no fire. The four-foot rocket climbed to 3,500 feet where it deployed a parachute and was recovered. These early launches were kept at a low altitude so the entire launch could be clearly seen.

Large Rocket Hanging Launch One   **06/20/95**

In this test, a large rocket was launched from a platform that was suspended from scaffolding. This launch was conducted with a nine-foot rocket powered by a motor a quarter of the power of our 60-mile sounding rocket. By now we had developed a pattern in our development process. Each new rocket would be launched from its platform suspended from scaffolding. When the scaffolding flight was flown to satisfaction, the rocket would be launched from balloons. Then we would increase the size of the rocket and start the process over again. The rocket flew to 9,500 feet.

Large Rocket Hanging Launch Two   **08/13/95**

This was the same test as the previous mission with a number of rocket improvements. This rocket carried an upgraded altimeter, parachute deployment flight controllers, and was spin-stabilized. The fins of the last rocket were damaged on landing. The new fins had Kevlar reinforcement. We used the fins as a baseball bat to verify their strength. The platform incorporated doors that blew open when the rocket motor was ignited. The rocket flew to 10,200 feet.

Large Rocket Rockoon Launch   **09/10/95**

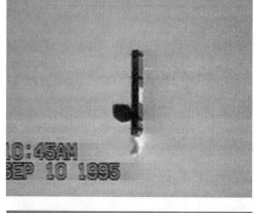

Three sets of balloons stacked in a 600-foot tall train were used to lift this rocket and launch box. The system was beginning to approach a scale that could achieve space flight. When the launch box reached 500 feet, the rocket was launched. The nine-foot rocket climbed cleanly away from the launch box. It continued up to 10,500 feet where it deployed a small drogue parachute. At 750 feet about the ground, the rocket deployed its main parachute. It landed softly a mile away from the lift-off point.

Vacuum Chamber Motor Ignition Test   10/08/95

A scale rocket motor with the identical chemical composition as the full size motor was test-fired multiple times in a vacuum chamber. The purpose was to test the motor-start characteristics at high altitude. The motor would ignite in vacuum, then it would blow off the end plate of the chamber to allow the exhaust plume of the rocket motor to escape. We were able to use the data to develop ignition techniques that would work in the vacuum at the edge of space.

Platform High Altitude Flight Attempt   06/12/96

This was the first attempt to fly the launch platform to high altitude. This was also our first use of high performance research balloons.

On the attempt we learned to appreciate the effects of the wind. Wind reduces the lift of a balloon, but this only has an impact on launch. Once the balloon is drifting with the wind, it climbs just fine. We tried to launch a six-balloon stack on a very windy morning. The wind flattened the balloons against the desert floor. The team carrying the launch box tried running with the wind in an attempt to get under the balloons. The Discovery Channel was out in the desert that day filming an unrelated event. They caught a very odd parade off in the distance.

The Discovery Channel narrator commented that it appeared that several people were being dragged across the desert by a bunch of balloons while holding a large coffin. This was the team's first TV appearance…ah, fame….

Twin Motor Rocket Launch
08/16/96

The supplier of our rocket motors went out of business, leaving us with a problem. No other manufacturer would build us a motor with the required specifications. To get around this problem, we designed a fourteen-foot rocket that used two available motors. We christened the rocket the "Twin Dilemma." A special fourteen-foot long launch box was built and suspended under a large "A" frame.

When the rocket was launched, one motor started late—a full two seconds late. The Twin Dilemma rose gracefully from the launch box and made a 45 degree arc. Now under the power of both motors, the rocket accelerated to just above Mach one. Now the real problem started. The motor that started on time also shut down on time, leaving the other motor running. The rocket tried to make another 45 degree turn, only this time upward.

The first turn was made at a relatively low velocity. This turn was at Mach one. After bowing for a moment, the rocket broke in half. The top half carrying the parachutes fell to the ground. The bottom half, carrying the parachute controllers, flew upward to over 10,000 feet before arcing over and slamming into the ground.

This inspired us to find another motor supplier who could make the larger rocket motor, and we abandoned the twin motor design.

Instrument Test Vehicle Rocket Launch 1 & 2   09/26/96

We developed a small rocket that would simulate the initial launch environment for the onboard instrumentation. This was used as a low cost method of testing rocket components. The first flight carried only the rocket's control system. The second flight carried a GPS system. The GPS was set to record the peak altitude. The GPS was able to get a fix and log the position.

Away 1 Instrument Balloon Flight   03/09/97

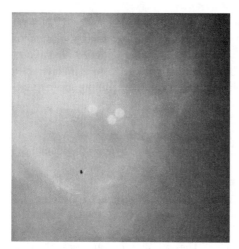

We needed the ability to test systems at high altitude. Before components could fly on major missions they could get a "shake down" on a simpler mission. To accomplish this, the "Away" series was started. Away 1 carried all the rocket launch platform electronics, but without the rocket launch box. Three balloons carried it aloft. The launch went well. It dispelled the dread created by the last launch attempt. At 60,000 feet, telemetry data showed that the battery levels were dropping fast. Within minutes the vehicle went silent. Tracking teams downrange were able to spot the balloon with binoculars. The altitude was an estimated 100,000 feet when they watched the balloons pop. After two weeks the search was called off. Away 1 was never found.

Away 2 Instrument Balloon Flight to over 100,000 feet   07/20/97

Away 2 was the second telemetry development flight. Away 2 was a much smaller vehicle than Away 1. It carried an altimeter, two cameras, a telemetry system, and a beacon.

It was primarily designed to test the flight computer and radio systems. Away 2 flew in excess of 110,000 feet. Its radio beacon failed on landing, resulting in the package not being found for two weeks. After being located, full forensics were carried out on the vehicle. The

lessons learned from this analysis had a huge impact on our system's designs. It was the mission that taught us how to fly in the upper atmosphere.

Space flight Rocket Ground Launch   08/16/97

The 'Space flight' rocket was launched from a launch platform that was suspended from scaffolding. This was the first test flight with the full power rocket motor. The rocket accelerates so fast at launch that the eye cannot actually see it. It was not even picked up on the video camera. This photograph was an exceptionally lucky shot by our photographer. The rocket shattered at 20,000 feet after reaching Mach three.

Comet 1 Instrument Balloon Flight to 5,000 feet   08/24/97

The Comet series of flights were designed to test the autonomous backup platform system. In the event of a telemetry failure of the primary command/control, the Comet system "safes" all launch systems, releases the balloons, and transmits a location radio beacon. In addition to the radio and release systems, the Comet carried a 35mm camera that took pictures every five minutes. Comets are about the size of a shoebox.

Balloons seem to attract wind. The wind was howling at seventeen knots on launch morning. The first balloon was blown to bits while trying to fill it with helium. The second balloon was stretching and distorting, looking more like various balloon animals than a weather balloon. At the time, we were measuring how much helium was in the balloon by hooking it to a fishing scale. We could see how much the balloon was lifting. With the balloon pulling in the wind, the scale was swinging from five to 45 pounds. It was impossible to tell how much helium we had used. We took a wild guess that the balloon was filled and sent the Comet skyward. It lifted to ten feet, then slammed down on the lakebed. It climbed up again, going slowly up, but racing downwind. The team leapt to the trucks and the chase began. Comets had the additional task of being a training vehicle for new team members.

The first Comet flew to 5,000 feet. The recovery team chased it as it flew. Comet 1 landed in the back of a truck speeding down the dry lakebed.

Second Sounding Rocket Suspended launch   01/19/98

This was the second test launch of our "Space flight" rocket. The launch was from a platform that was suspended from scaffolding. This was the second test flight with the full power motor. The rocket flew to 35,000 feet at a velocity of Mach three. At this point in the flight, the dynamic loads shattered the rocket into thousands of small pieces. This rocket had a paper/epoxy airframe. It was designed to operate in the near vacuum environment of 100,000 feet. The rocket breakup showed that if the rocket were fired inadvertently downward, it would not reach the ground. The rocket would breakup up when it entered the lower atmosphere. In addition to testing the rocket, this flight was also the final validation flight for the space attempt's launch box design.

Comet 1b Instrument Balloon Flight   04/09/98

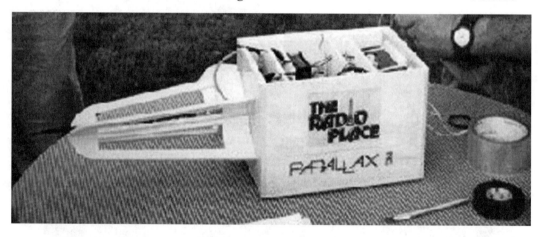

This was the second Comet development flight. It was lifted by a surplus Navy balloon. The Comet was flown to 80,000 feet. Improvements to the "B" model included a more powerful radio and sliding racks for equipment mounting.

Comet 1c Instrument Balloon Flight   04/18/98

This was the third Comet development flight. In this test, the Comet was held on a tether at 800 feet while it performed its function. This allowed observation from the ground of the balloon separation (FTS—Flight Termination System). This Comet also carried a six-volt solar panel for recharging batteries in flight.

Comet 1d Instrument Balloon Flight   **05/24/98**

This was the fourth Comet development flight. In this test, the Comet was flown to 67,000 feet. The Comet carried a 35mm camera. This Comet featured a more powerful computer, an improved antenna, and an extensive software upgrade. The "D" model would be the model used as a support package on future Away and space flight missions.

This was the last full Comet to fly independently. Future Comets were integrated as subsystems in larger vehicles. Mini Comets took over the training role.

Vee Balloon Tether Flight   **06/16/98**

This was the first flight test of our vee-shaped balloon system. This was the basis of the first stage airship. Rows of three-foot diameter balloons were attached to a lightweight frame. The sixteen-foot long vehicle was flown on a tether to 500 hundred feet. Basic stability and drag tests were conducted. Wind-handling was an amazing improvement over free balloons.

Away 3 Instrument Balloon Flight   **07/05/98**

This was a launch platform test flight. It contained the entire command/control system, the Comet system, live video platform inspection system, and a full internal system monitoring downlink.

It was carried aloft by three weather balloons. The platform was flown to 36,000 feet. This would be the platform electronic configuration for space launch. The platform landed in a rice field. It continued to transmit until the recovery team located it in spite of being underwater.

Rocket Telemetry Test Flight

A low power and low cost version of our "Space flight" sounding rocket was used for flight telemetry tests for the rockets avionics. This rocket used a motor a quarter the size of the one used for space flight. It was launched from a standard launch pad instead of the enclosed launch platform. The rocket was flown to 10,000 feet. This flight verified that we could GPS position reports from the rocket in flight.

Away 4 Instrument Balloon Flight to 108,000 feet   01/20/99

Away 4 was a balloon flight for testing rocket equipment. The "Space flight" rocket's GPS/telemetry system was carried to 108,000 feet to test it at high altitude. Away 4 encountered a reverse jet stream above the regular jet stream. The balloon release mechanism failed, leaving Away 4 stuck at the edge of space. The reverse jet stream carried the package 600 miles out over the Pacific Ocean. Away 4 carried a live video transmitter. The last view from the vehicle showed a vast blue ocean. The scene faded to the snow of static. Away 4 was lost; however it was an excellent long distance test of the system.

Rocket Gas Chute Ejection System Flight   **02/14/99**

The low power test rocket was used to test the new cold gas parachute ejection system. This system replaced a pyrotechnic parachute deployment system. The rocket was the low cost version of the three-inch diameter "Space flight" rocket.

Vee Airship Powered Scale Model Flight   **02/21/99**

A five-foot scale model of the Vee-shaped airship was flown on a series of free flights. A single propeller powered the scale airship. Control surface and general stability tests were conducted.

Mylar Balloon Prototype Flight   **03/05/99**

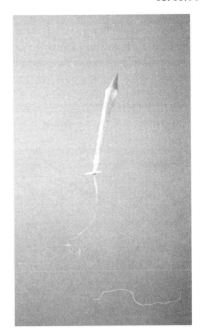

This was a test flight of our in-house manufactured balloon. This balloon has a long and narrow configuration. This allows for much faster climb rates. The balloon used a Mini Comet to provide telemetry. The Mini Comet also controlled a vent panel to release the helium for landing. We called these fast-climb balloons "Rabbits."

Rockoon launch to 72,233 feet   **05/23/99**

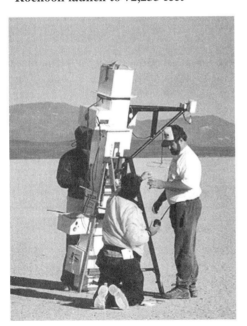

This was a full rocket launched from a balloon-array at high altitude. The launch platform was carried aloft to 29,000 feet. The rocket was then launched to an altitude of 72,233 feet. It took 40 team members to accomplish this flight. Reporters and news crews from around the world joined us on the lakebed.

Half Scale Flyback Platform Flights   06/29/99

This test series was designed to explore the potential of adding aerodynamic surfaces to the launch platform. This would allow the launch platform to fly back to the lift-off site after launching the rocket. The purpose of this is to reduce the ground recovery time of the launch platform. In this series of flights, half-scale models were flow from a bridge to determine basic flight characteristics. The first dozen tests were free flights. The second dozen were computer controlled. Four vehicles were built.

Half Scale Flyback Platform Airplane Launched Flights   07/10/99

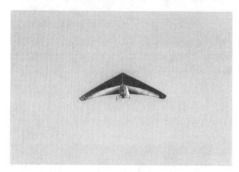 

The half-scale flyback platform was mounted under an ultralight aircraft. The platform was released when the aircraft reached 1,200 feet. The platform then flew to the ground. The platform proved to be very pitch sensitive when carrying high loads. Normally, a platform descending from 100,000 feet would be very light. However, the platform must be able to fly back with the rocket still on board. Nine drop flights were performed.

Full Scale Fly back Platform Flight   07/17/99

This test flight was a continuation of the "fly back" tests with the full-scale launch platform. The flyback launch platform was carried aloft by balloon and released. The balloons were on tether and rose to 300 feet before the platforms were released. They were then flow down by remote control. There were two models of the full-scale flyback platforms built, one with foam wings and one with nylon wings. With the development of the Dark Sky Station, the flyback platform was no longer needed and the development was ended.

Mini-Comet Flight to 100,000 feet   08/13/99

        The Mini Comet is a small low cost version of the Comet series instrument package. The purpose of the Mini Comet is to provide training flights for the vehicle ground recovery teams. It consists of a computer, a camera, radio beacon, and a balloon release system all enclosed in a foam housing. The Mini Comet is an excellent camera platform for high altitude photography.

First Launch of MicroSat Launcher Upper Stage   09/18/99

This was a basic configuration test flight of the MicroSat Launcher rocket. The rocket was flown to 10,000 feet to a velocity of Mach 1.1 and successfully recovered. The MicroSat Launcher (ML) was designed from the ground up to be launched from balloon. Its wedge fins were optimized for Mach five at 200,000 feet. Instead of a conventional launch pad, the ML was launched from a set of four rails as it would be launched from a balloon. This was the "Block One" configuration. It has a fiberglass airframe, a carbon fiber nosecone, and fins.

Scale Advanced Platform Tether Test Flight   **10/15/00**

This was a tether flight of a one-quarter-scale frame platform. The platform was eight feet across and weighed three pounds. Twelve balloons lifted it. It carried a quarter-scale launch box. Balloons were used to lift the tether line. This was to eliminate any pull from the tether line. This made the lift-off and climb more realistic. The platform was flown to 600 feet.

Advanced Platform Test Flight   **01/29/00**

This was the first test flight of the Advanced Platform structure. This structure is twenty feet wide, five feet tall, and weights four pounds. An array of nine balloons was attached to the outer rim mounting points, and a seventeen-pound launch platform was mounted in the middle. The unit was flown to 15,000 feet in an initial configuration flight. The platform proved to be extremely stable and exhibited excellent liftoff characteristics. This platform is half-scale to the one that could carry the ML rocket to launch altitude.

Space flight Attempt   **03/25/00**

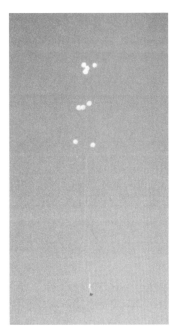

This was an attempt to launch a rocket from the balloon platform to 57.5 miles. Ten balloons in all lifted the rocket in its launch box. The height from the top of the balloons to the bottom of the launch box was 800 feet. To make the launch easier, a powered winch was used to control the balloon stack before launch. Launching the big stack was still an extreme challenge. We managed to get it into the air, but just barely.

Midway through the flight, the onboard GPS was not consistently holding a lock on the GPS satellites. This caused the position data to be intermittent. This constituted an abort condition and the flight was terminated. This was to be the last flight using the tall stack balloon approach.

Cylindrical Rabbit Balloon Test Flight 1   **05/28/00**

We are developing cylindrical Mylar balloons for high speed climbs. These flights consisted of a tall, twenty-foot balloon and a Comet as a flight control system. In addition to testing the new balloon design, the flight was another opportunity to test the Comet system that is used as a backup flight termination system for the space flight missions. The mission consisted of a vertical climb to 1000 feet, at which time the Comet deflated the balloon. The Comet then used the deflated balloon as a streamer to slow descent. The cylindrical balloon showed a fast climb rate even with low positive lift.

Cylindrical Rabbit Balloon Test flight 2   05/28/00

This was the same as above except the flight was extended to 5000 feet. The system was ready to go within minutes of each flight. The flight was ended by flipping the balloon upside down, allowing the helium to rush out of a hole at the bottom of the balloon. A line was attached to the top of the balloon from the Mini Comet. The computer released the Mini Comet from the balloon. As it fell, it would pull the top of the balloon down, turning it over. It showed the reusability of the balloon and the gas venting system.

MicroSat Launcher Upper Stage, Block II launch   06/15/00

This was the second generation of the MicroSat Launcher Rocket. The primary upgrades were a lighter, all carbon fiber airframe, a Kevlar nose cone, and the full space flight command/control and GPS tracking system. The ML carried a ten-pound steel weight to simulate a payload. The rocket was launched from the full three-axis gimbal system as it would from the balloon platform. The rocket motor was slightly larger than the one used on the prototype ML launch. The rocket climbed to the predicted altitude of 16,400 feet. It then deployed its parachute and descended to Earth. The rocket landed softly and was ready to fly again without any repairs required.

Away 6 Instrumentation Test Flight   10/13/00

This was an instrumentation development flight. It carried the full space flight electronics set. The thirty-three pound instrument package was carried aloft by four weather balloons. Away 6's payload included two 35mm cameras, live video transmitter system and a 8mm movie camera. Away 6 flew to 92,217 feet.

On descent we tracked the package with GPS down to ten thousand feet. The valley that it landed in was only forty miles away, but it was in the high Sierra mountains and nearly inaccessible. We mounted a recovery expedition the next weekend. When the search team arrived in the valley they found it was completely covered in snow. Two days were spent trudging through search lines looking for a white vehicle under two feet of pristine white snow. The vehicle was never found.

Away 7 Rabbit Expansion Cell Test Flight   10/14/00

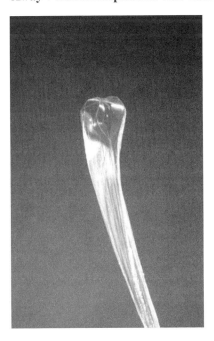

The development of our own Mylar balloons had progressed enough for a mid-altitude test flight. A major step toward long duration balloon flight was accomplished. In flight, three additional

Mylar balloons were inflated. This was the first step toward in flight balloon replacement. The vehicle was flown to 24,000 feet. The secondary balloons were stored in side-mounted containers during the launch. They were successfully inflated with helium from the primary balloon in flight.

StratoStar Frame Tether Flight Test   **11/07/00**

This StratoStar was another platform design. Instead of a true platform, StratoStar was a balloon spacer. It held the balloons 30 feet apart. The rocket launch rail was suspended below. The test was a success. However, it showed that the spacer fame would only be a slight benefit over balloons alone. The benefits were not enough to justify the weight and complexity.

Cathedral Frame Tether Flight Test   **11/07/00**

This was a launch and stability test of a balloon platform design. This was a full-scale version of the earlier advanced platform design. The Cathedral platform was amazingly light for its size. Like the Stratostar, it was designed to lift the MicroSat Launcher Rocket to 100,000 feet. The difficulty of the Cathedral design was its complexity. It took over nine hours to assemble after unpacking. Field assembly would take even longer. The Cathedral used hundreds of feet of tension lines. It was the precise adjusting of tension lines that took the bulk of the setup time.

The Cathedral frame was very light; its total flight weight was 24 pounds. It was also a beautiful structure. Its downfall was deployment. It was impossible to deploy in the field. There was an intense rivalry between the designs of the Cathedral and the StratoStar. It became known in by the launch teams as the "Frame Wars."

Dark Sky Station Tether Flight Test   **12/09/00**

The Dark Sky Station (DSS) was the next platform design to be developed. It was a direct leap to the general shape of the eventual ATO station. The DSS was a truss structure made from carbon fiber rods. There were four trusses forming a giant "X" eighteen feet across. The DSS was repeatedly launched and flown to 150 feet. Several tether lines attached to the ground limited the maximum altitude. The purpose of the test was to observe launch-handling characteristics and to train launch personnel. The DSS was a dramatic improvement over all the prior platform designs. The Frame Wars were over.

Dark Sky Station Tether Flight Test (five arm)   **12/23/00**

This larger vehicle was beginning to look like the future city in the sky. This DSS had an extra arm and a central hub holding it all together. The five-arm vehicle has more lift capacity and stability. The platform flew with a command/control, GPS tracking, and a Comet backup system. The DSS was launched repeatedly on tether. The first ten launches were used to determine basic handling. The next twenty launches were just for the sheer joy of watching her fly.

Mesospheric Explorer Prototype Flight   **05/19/01**

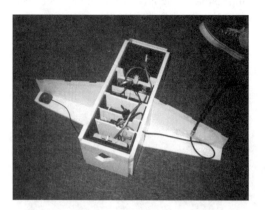

The Mesospheric Explorer is a small, high lift balloon/instrument payload system. It is designed to climb to extreme altitudes (over 130,000 feet). This flight was the initial system test flight. The climb rate on this first test flight exceeded 1,300 feet per minute. At 30,000 feet, the vehicle experienced a complete system failure. Three independent and separate systems failed simultaneously. The Mesospheric Explorer was never seen again. To this day its loss remains a mystery.

Dark Sky Station Free Flight   **05/19/01**

The first free flight of the Dark Sky Station was conducted in the Black Rock Desert in Nevada. The sixty-five pound craft was successfully launched after an hour hold due to surface winds. The DSS was flown 73 miles downrange where the balloons were released and the parachute deployed. The vehicle was recovered four hours later. This flight opened the door to the launching of large structures for high altitude work.

Large Dark Sky Station Tethered Flight 1   8/26/01

The 57-foot diameter DSS was test-flown attached to ground anchor lines. The basic stability was verified. Several launch techniques were tried in order to figure out how to get this monster off the ground. The station was very stable in free flight, but unstable on tether. We found out very quickly that our parking lot wasn't big enough for this giant.

Large Dark Sky Station Tethered Flight 2   9/23/01

The DSS was flown higher and in stronger wind conditions. It has excellent free flight handling in winds up to twelve knots. However, when it reached the end of the tether, there were problems. When the tether went taut, the station turned nearly ninety degrees like a kite. This dramatically increased the structural loads. One of the arms was damaged, and one broke away completely. The tether was not going to be part of an actual mission. It was only part of the test to keep the DSS from flying away. We made the decision to move forward with the free flight of the vehicle.

Mini Rack Instrument Balloon Flight   10/06/01

The mini rack was both a test of a new structure and a data gatherer. The mini rack was a carbon fiber two-shelf platform. One balloon carried the vehicle to 108,000 feet. It had an average climb rate of 1,200 feet per minute. This vehicle was used to measure the winds aloft prior to any large platform flight. This particular flight was used to map winds aloft for the large Dark Sky Station flight.

Large Heavy Lift Dark Sky Station Flight   10/06/01

This was the first free flight of the heavy lift Dark Sky Station. This large DSS is designed to carry the JPA space flight rocket to 100,000 feet for launch. On this flight, there was a rocket launch box. However, it did not carry a rocket.

The team began launch prep at 3:00 a.m. for a launch at 9:00 a.m. Balloon teams began filling the ten balloons it would take to lift the station. The big vehicle was our most complex to date. It had a wireless network that could release an individual balloon or all the balloons at once or even various patterns of balloons. The wireless system was backed up by hardwire connections. There were two command/control systems running in parallel, two GPS tracking units, two parachute cannons, two still cameras, a camcorder, and a transmitting video system that provided views from five cameras. A Comet was also carried. The Comet could takeover and terminate the mission if there was a problem

with the primary and backup control systems.

When dawn arrived, the light night breeze stopped, leaving a prefect morning for launch. The large balloons above the station's arms were perfectly still. After three hours of going through the checklist it was time for launch. Under each arm was a member of the launch team holding the station down. On command, they all let go. The station leaped into the air climbing over 1000 feet per minute.

Moments into the flight, the crowd below could see that something was wrong. The balloons were rocking back and forth. As DSS 2 climbed, the rocking got more severe. The balloons were rocking so much they were actually swinging under the station. Not only were they swinging violently, but the were beginning to swing in sync with each other. At 3,000 feet, the station began to tear itself apart.

The ripping and snapping of the carbon framework could be heard clearly from the ground. Even though the structure was a mess, the electronic systems were still working. Mission control sent the command to release the balloons. The wreckage fell fairly slowly, around twenty miles per hour. This is the flight that hit with that crunch like breakfast cereal, a hundred yards from the launch point.

We had views from five video cameras on board and four on the ground. This gave us the opportunity to really investigate what happened. Our research showed that when traveling above a certain speed, a low-pressure zone forms beneath the balloon. Low-pressure zones between different balloons can interact if the balloons are close together.

This phenomenon can make the balloon unstable in some configurations. However, it can be used to make balloons very stable in others. The problem that crashed Dark Sky Station 2 became the key to the Tandem twin balloon airship.

Away 13 Balloon instrumentation flight   **03/09/02**

Away 13 had several purposes. It was used to test the new command control system of the Dark Sky Station. An improved primary balloon line pyrotechnic was also tested. Away 13 used the JPA carbon rack system as the structure. It weighed nine pounds and flew under a single balloon with 36 pounds of lift. The testing required a minimum altitude of 65,000 feet. Away 13 was flown to 72,000 feet and was recovered 55 minutes after landing. The flight was a great confidence builder for the team after the crash of DSS 2.

This was inaugural flight of the PongSat student payload program. Fifteen student experiments were flown.

Away 14 Weather Research Balloon Flight   **3/23/02**

JP Aerospace conducted the first flight and opened the new Oklahoma spaceport. Two high altitude balloon flights were flown. The first, Away 14, carried meteorological instruments from the University of Oklahoma as well as PongSats. The local fire department and several elementary school students assembled the PongSats. Away 14 flew to 103,000 feet.

Away 15 Paper Airplane Drop   **3/24/02**

Away 15 was the second mission flown for the opening of the Oklahoma spaceport. This was an educational mission. The vehicle carried over 600 paper airplanes made by the school children of Oklahoma. At 90,000 feet, the paper airplanes were deployed. For the next several months, people all over the state found the paper airplanes. The landing sites of the airplanes were tracked by the State of Oklahoma.

Away 9 Small High Rack Configuration Test Flight   **5/11/02**

The new vehicles we built for the Oklahoma flights delayed the Away 9 mission. This flight introduced our new standard rack configuration. This standard combination of systems allows a shorter lead time before flights and a simpler, lower cost operation. Away 9 carried 77 PongSat student experiments. Away 9 flew to 90,000 feet. The recovery teams really moved on this flight. They were at the landing site within twenty minutes of touchdown.

Away 16 High Climb Rate Balloon   05/11/02

This balloon was designed for rapid climbs. Like the Rabbit balloons, it was a long narrow cylinder. The balloon was made out of ripstop nylon coated with Teflon. At the base of the balloon were stabilizer fins and an instrument rack.

This launch was exciting. The wind picked up to eighteen knots while the balloon was being filled. This mission taught us a lot about managing "launch confusion." While wrestling with the balloon in the wind, the checklist process was compromised. The tracking system was turned off without noting it in the checklist. The balloon was getting hard to handle and the crew running mission control jumped in to help. With no one at mission control, the tracking system going silent was not noticed. An hour after liftoff, the backup Comet unit released the helium and Away 16 landed. We had beacons giving us direction, but no position on the package. The vehicle was never found.

Away 19 PongSat High Altitude Balloon Flight   10/05/02

As part of opening the West Texas Spaceport, Away 19 carried 62 PongSat student experiments to 100,000 feet. This flight also provided a test of our miniature "TinyTrak" GPS tracking system. Eighteen-knot winds blew throughout the day. The wind blew apart three balloons before the team finally got the vehicle in the air.

Ascender 20 Tether Test   10/06/02

Ascender 20 was built as a miniature test bed. It was used as a research tool for the Vee airship. The airship is twenty feet long with three-foot diameter envelopes. Carbon fiber tubes formed the keel. This airship started its life with clear plastic envelopes. After an initial series of tests, the plastic envelopes were replaced with nylon outer envelopes and Mylar inner helium cells. Flight tests on a tether line were conducted at the West Texas Spaceport in Fort Stockton, Texas. The tether tests were used to determine basic stability and handling. We flew the airship in the rain and wind. It was a great shakedown for the vehicle and gave the team a good soaking.

MicroSat Launcher Rocket Launch   10/05/02

This was the third flight of the MicroSat Launcher Rocket. It was the first rocket launch from the new West Texas Spaceport. Over 50 VIPs from West Texas, including senators, congressmen, judges, and reporters were on hand to watch the inaugural launch. The winds at launch time were at 24 knots. One more knot and the launch would be scrubbed. The team wasn't worried about the launch. The ML could handle much higher wind. What gave the launch director (the author) a pit in his stomach was the direction. The wind was blowing directly from the launch pad toward the VIPs. Visions of the rocket floating down under parachute into the crowd filled my head. Ten minutes before liftoff we made a decision. We would angle the rocket downrange. This would put the rocket many miles away when the parachute deployed. The drawback was the destruction of the rocket. The rocket lifted off on a 14,000 n/s motor and flew to 20,000 feet. Flying straight up the rocket slowed down to a near stop before the parachute was ejected. This is how we like it. The parachute unfolds in very slow air. When launched at an angle the rocket makes an arc like an artillery shell. It slows down very little. When the parachute deploys, the rocket is still moving very fast. The ML would be moving at Mach one. At this speed, the loads are more than the structure can handle. We thought it was better to break the rocket downrange out of sight rather then scare the crowd by floating down in the middle of them. Floating down by the spectators would have been safe; however, it could have given everyone a fright. It probably would not leave a good impression for opening day. After the countdown, the ML leaped into the sky. It put on a good show. When we got to the landing site, we where shocked to find a completely intact rocket. There was some airframe damage and the parachute had a big tear, but that was it. The ML is a tough bird.

Away 17 Instrumentation Balloon Flight   **10/18/02**

Away 17 continued the school of operations at the edge of space. Each of these runs to the top of the atmosphere builds not only experience, but helps to work out all the little fiddly bits. It's amazing to me how it's the small details worked out on Away missions that make the big flights a success. Away 17 was flown to 94,795 feet. The flight tested a critical upgrade to the flight command control system. The vehicle also carried 119 student experiments. On board video recorded a daytime sighting of the planet Venus.

Away 18 In-House Manufactured Mylar Balloon Flight   **10/18/02**

Away 18 tested the next version of the in-house manufactured Mylar balloon. The flight carried a GPS tracking system and a balloon separation system. At 41,352 feet, the balloon popped and package started down. It landed only a mile away. When we inspected the balloon, we found that all the tape had separated from the Mylar. The balloon is made of Mylar panels taped together. We discovered the tape we used completely lost all stickiness at 50 degrees below zero. All the tape simply fell off the Mylar. What was once a balloon was now just a bunch of strips of Mylar.

When we came back from the desert, we conducted tape freezing tests. We bought rolls of every kind of tape we could get our hands on. We would freeze-test samples down to –119 degrees F and see how they would hold up. We finally found one that would hold up, and we started building the Away 20 balloon.

Ascender 20　　　　　　　　　　　　　　　　　　　　　　　　**11/11/02**

The small Ascender was flown inside a large horse arena. We flew the Ascender through powered time trials. The Ascender was flown across a 100-yard course. The "pilots" flew it with model airplane radios. It took a dozen runs to get the feel of the airship. It's a good thing airships just bounce when you fly them into walls. It was powered by two electric motors, each spinning twelve inch propellers. We were able to change the angle of the vee arms and compare performance. After the data was collected, we flew the vehicle outside for its first free flight.

Away 20 Large In-House Manufactured Mylar Balloon Flight　　　　**12/8/02**

This balloon was a monster at 80 feet long and fifteen feet in diameter. Away 20 included a GPS tracking system, a balloon separation system, and a new ultra light mini beacon. Away 20 flew to 91,000 feet with climb rates of over 1,400 feet per minute before landing 59 miles away.

Away 21 Instrumentation Balloon Test Flight　　　　　　　　**12/8/02**

This mission was loaded. Away 21 was a High Rack flight balloon flight with all the bells and whistles and over 300 PongSats. The mission tested a new ultralight tracking system and a new balloon release mechanism. The winds were dead calm at liftoff but fast at altitude. The vehicle climbed to 94,100 feet and landed 63 miles down range.

Ascender 90 Lift Test   **02/22/03**

Ascender 90 was our first large vee-shaped airship. It was 93 feet long and each arm was fifteen feet in diameter. Ascender 90 had all the elements of the ATO stage one airship: outer shell, Mylar inner cells, carbon fiber keel, and a helium management system.

We filled the airship with helium and floated it to ten feet off the ground. We assembled the airship in the parking lot. The test was done at night for calm winds and also so we wouldn't get run over. The vehicle was so big it brushed the buildings on both sides of the parking lot.

Ascender 90 Full Instrumentation Test   **04/25/03**

The Ascender 90 uses an onboard wireless network. The network manages the helium pumps, valves, vents, navigation, and propulsion motors. All the systems were operated as if the airship were in flight. This test shook down the entire system.

Away 22 High Altitude Propeller Test Flight   05/21/03

The Away 22 mission tested the performance of the Ascender's carbon fiber propeller at high altitude. An array of six balloons carried the propeller and electric motor to 80,000 feet. After the tests,

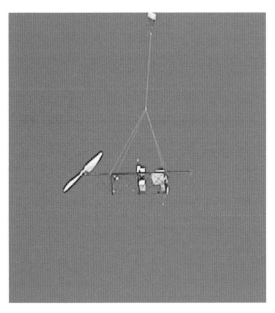
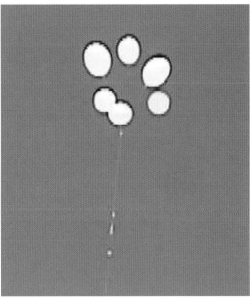

the vehicle floated to the ground on three balloons. The vehicle landed hard and the motor and propeller broke off. Away 22 was now light enough to climb back up again. The telemetry system was also damaged so the vehicle drifted silently away. The team had no idea this had happened until they reached the recovery site. There was the electric motor with the propeller still attached along with drag marks that just vanished. It was in a very remote area on a mountaintop. It was unlikely that anyone got there first. Two weeks later, the rest of Away 22 was found. A power line crew came across it 200 miles further downrange. Unfortunately, the camcorder, digital camera, and parachute had been stolen. We still had the transmitted video and data. We would still love to get the tape and camera memory cards back, no questions asked. The flight showed the propeller's design has excellent performance at extreme altitudes. The mission also tested the complete Ascender electronics system. The flight also carried 32 student PongSat experiments.

Hybrid Rocket Flight   06/09/03

This demonstration rocket flight was conducted for a West Texas teacher's conference on PongSats. We took this opportunity to conduct out first launch using a hybrid rocket motor. A small, four-foot rocket was used. The launch and recovery were successful.

Ascender 90 Float Test   06/22/03

The Ascender was run through its paces to test propulsion and helium management systems. We finally acquired a hangar large enough not only to float the airship, but also to fly it around a bit too. The main engines were installed and the propellers spun up. Pitch and roll and turn maneuvers were tested, and short runs across the hanger were made.

Large Hybrid Rocket Flight   09/27/03

This rocket flew on a 9,000 n/s hybrid rocket motor. This flight was used to evaluate the use of hybrid rocket motors in the ML Launch program. This twelve-foot long vehicle was flown to 11,000 feet. Hybrid motors add to the complexity of the launch. However, after the initial investment in launch equipment, they are very economical to use.

Hybrids are also at the core of some of the orbital airship propulsion system concepts. This flight was an opportunity of adding to our experience with this technology.

Away 23 Instrumentation Balloon Test Flight   10/25/03

Away 23 was a training mission for mission control operators. It was to prepare for Ascender flights. This flight used a polyethylene zero-pressure balloon. When we release the vehicle it didn't go up. It just sat there four feet off the ground. We pulled the balloon down and discovered the problem. There was a one-inch hole in the side of the balloon. We slapped a tape patch on it, pumped more helium in, and we were back in business.

In addition to our own equipment, 174 PongSats student experiments were also on board. The balloon didn't seem to mind the patch. The balloon reached an altitude of 100,935 feet.

Away 24 Ascender Network Test Balloon Flight to 100,700 feet.   1/17/04

On launch, the vehicle zoomed up at 1000 feet per minute and continued at that rate to peak altitude. The flight was excellent. The only glitches were in the imaging systems. Both the video and the still camera system had power failures shortly into the flight. This mission tested systems designed for the Ascender airship. Sixty PongSat experiments were on board. The PongSat experiments ranged from materials testing to measuring the effectiveness of a solar panel at high altitude.

The burst of the balloon was observed by the team through a telescope. Retrieval was its own adventure. The vehicle landed far from any accessible road. A recovery team made a three hour

bouncing, sliding off-road ride followed by a nine hour hike guided by GPS. After experiencing darkness, below zero temperatures, mud, snow, and two wolf packs, Away 24 was recovered. The event has become known within the team as "Recovery with Wolves."

Ascender 175 Float Test   1/29/2004

The 175-foot Ascender was filled with helium and floated in a large hangar. This was the first time the airship could be fully assembled. The large vee-shaped envelope had to be floated over, then landed on the external keel. The propellers were spun up and sub-system tests were performed.

All was not well. The inner gas cells were filled with helium and the giant envelope was floating ten feet above the floor. As it was walked back over the keel structure, everyone could tell something was wrong. As the one was lowered over the other it became obvious. To our horror they didn't match. The ends of the vee were 30 feet wider on the envelope than the vee of the keel. Quick measurement showed that the angle of the vee of the envelope was off. We had sewn the 90-foot Ascender ourselves. The Ascender 175 was just too big, and we contracted another company to make the envelope. The company didn't have a building big enough to inflate it to the check the measurement. It was constructed in the winter, and inflating it outside was out of the question. This left us with the surprise on assembly day. We tried bending the legs of the vee in to match the keel. The force required to bend it even a few feet was too much for five people to hold. Thirty feet was out of the question. The keel needed to be rebuilt and widened.

Away 25 High Altitude Test Flight   4/3/2004

This flight used three zero pressure polyethylene balloons. These large plastic balloons will be used in groups on future Dark Sky Station missions. The High Rack itself was completely reconfigured. The new rack is lighter and shorter, with wider shelves. The parachute was not held up in line with the balloons as in earlier flights. The parachute was stowed on one of the stabilizer fins and deployed when the balloons were released. Also new was a central power system. The central power

system cuts down on the need for individual batteries on each device. This was the first flight that we used lithium polymer rechargeable batteries. The main goal for the flight was to test electronics for the Ascender high altitude airship.

The launch was conducted from a flat outcrop that overlooks the lakebed 300 feet above the desert floor in Black Rock, Nevada. On launch, the vehicle climbed up at 600 feet per minute. Near peak altitude it climbed a little faster at 800 feet per minute. During the flight, a helium pump designed for the Ascender airship was tested. This mission-tested system was designed for the Ascender airship and Dark Sky Station platform. Away 25 was heavier than most High Racks and climbed slower. The flight time was nearly two hours longer than most High Rack missions. The 343 PongSat experiments on board got the best exposure to the edge of space than any other mission so far.

At 96,050 feet, one of the balloons burst. Away 25 began to slowly descend. At 20,000 feet, the vehicle was over a valley between two mountain ranges. The command was sent to release the balloons and deploy the parachute.

MicroSat Launcher Rocket Satellite Deployment System Test Flight   **4/17/2004**

The MicroSat Launcher rocket once again thundered off the pad in West Texas. This time the ML carried a small satellite as part of a deployment system test. There is a backlog of satellites that have been built by university students but have no rides to space. Most of these will sit on a shelf forever, never having a chance to fly. It is better to have a short ride than none at all. The satellite, made at Stanford University, was ejected at 10,000 feet and descended by parachute. This mission was flown for the Air Force as part of a small launcher development program.

Mini Comet 3   **11/3/2004**

The Mini Comet is a 15-ounce instrument package. This was a training mission for some new team members. Mini Comets give new folks a chance to learn every aspect of a mission on a small scale. The team built the balloon, control system, camera actuator, beacon system, and housing. The Mini Comet flew to 20,000 feet and landed fifteen miles down range. The Mini Comet landed well offroad. It gave the new team members a taste of recovery operations.

Away 26 High Altitude Research Mission   5/21/2005

This mission accomplished several goals. It was our first flight of a spread spectrum telemetry system. And our new mission control team got their first flight experience at the console. Away 26 carried 202 PongSats.

Away 28 Roller System   8/15/2005

Away 28 is an odd looking vehicle, even by our standards. It is a high rack, with a large drum attached. A polyethylene balloon is wrapped around the drum before flight. The first few feet of the balloon is unrolled and filled with helium. The vehicle is then launched. As it climbs, the helium expands. As the balloon expands, more balloon is unrolled from the drum.

Its simplicity is deceptive. It explores a critical technology for the Dark Sky Station; the ability to manage and maintain the balloon in flight is the key to long duration balloon flights.

Away 27 High Altitude Research Mission   6/2/2006

Canceling missions due to high winds is the largest expense of the program. We needed an all-wind launch system. Expanding on the work of Dr. David Rust and Dr. Thomas C. Marshall, we designed a balloon bag to contain the balloon on the ground before the launch. Balloon bags have been used extensively in the past for small balloons. Large balloons pose a greater challenge. After six months of designing, "in the park" tests, and sewing, we had a launch system that could launch twenty foot diameter balloons in high winds. Away 27 was the first full-sized launch using this system.

The balloon bags allowed us to launch on much higher winds. It also allowed us to discover the next wind launch problem: a crosswind 25 feet above the ground. At launch, the crosswind took the balloons to the side of the Away 27 instead of above it. The launch team tried to get the vehicle under the balloons. They got close, but not close enough. When Away 27 pulled away from us, the balloons were still lifting at an angle. As it swung under the balloons, it hit the ground and tipped onto its side. The whole thing was over in less than a second, and Away 27 climbed away.

We picked ourselves up and rushed over to mission control to check the damage. Telemetry was a mess. The primary controller and one tracking system were down. The backup controller and beacon were fine. The secondary tracking system was not updating positions. The mission control team started to see what they could do with our injured patient.

Fifteen minutes into the flight, we reached a go/no-go point. Away 27 suffered too much damage to continue and we made the decision to abort. At 20,000 feet, the balloons were released, and Away 27 started down. The team was able to follow the vehicle down with binoculars. It was very apparent that the parachute did not deploy. From the onboard cameras afterwards, we could see that the drogue parachute was pulled out and tangled from the launch bounce.

The vehicle was a mess. The decks are designed to collapse on landing to absorb energy. Everything is packed in foam boxes that can take a bit of abuse. The structure was destroyed, but no systems were damaged from the ground impact. Unfortunately, the PongSats took most of the damage. Dozens were dented and twenty were broken open. The bag holding the PongSats had torn. PongSat were scattered across the desert. Of the 580 PongSats, all but three were found.

Scale Tandem Tether Launch   7/15/2006

The Tandem is a twin balloon airship designed to fly above 100,000 feet. It would use dual launch bags for launch. This would allow the airship to be launched in high winds. A three-quarter scale model was built to test out the launch technique and train the team. We launched the scale Tandem several times. It was attached to a 50-foot tether line to keep it from flying away.

Away 29 High Altitude Research Mission   11/4/2006

This mission was the second mission using launch bags. The balloons were contained in giant bags on the ground. At launch, long Velcro "zippers" were pulled, releasing them. The bag system lets us launch in high winds.

The launch bags worked perfectly. Away 29 began its climb to the edge of space. At 95,100 feet, one of the balloons popped. Away 29 descended on one balloon all the way to landing.

Away 29 flew not only for research, but for marketing. The vehicle carried 32 company logos and advertisements. Six cameras on booms took 2400 pictures of the promotions in flight.

Away 30 High Altitude Research Mission   11/4/2006

Away 30 was launched one hour after Away 29. The dual vehicles were designed to give mission control experience managing multiple vehicles. Away 30 carried 580 PongSats. It was also used to further test the new spread spectrum telemetry system. Away 30 was also launched with balloon bags.

Moments after liftoff, we could tell something was wrong. The vehicle was climbing too slowly. Away 30 would not reach the goal altitude before it flew beyond the recovery zone. We decided to end the flight. At 17,600 feet the command was sent to release the balloons. Away 30 landed under parachute six miles away from the launch site.

As I write, the team is working on Away 34 and there is a big airship out back waiting to fly. So as they say in the serials, *to be continued....*

INDEX

2001 A Space Odyssey 9, 50

A

Ad Astra rocket company 116
Aerobee sounding rocket 57
Aerospike engine 117
Aerovironment 48
Apollo balloon 62
Ariane 10
Arlandes, Marquis Francois d' 53
Armstrong, Neil 10
ATO 12-14, 16, 18-19, 21, 28-30, 32-33, 37,
39-41, 46, 51, 55, 60, 67, 69, 75, 77-78, 82,
89, 92, 94, 99, 101, 103, 109, 111-112, 117,
122, 129-132, 137-138, 140-141, 145-147,
151, 159-161, 179, 190

B

Beachcraft Starship 139
Bergerac, Cyrano de 67, 85
Bespin 68
Bigelow, Robert 62
Birnbaum, Ernest 54
Blackbird 47
Blue jets 40, 42
Boeing Dreamliner 139
Bova, Ben 69
Breitling Orbiter 55, 90

C

Cardiff Centre for Astrobiology 44-45
CEV 9
Coimbra, Portugal 52
Columbia space shuttle 9, 19
Conan Doyle, Sir Arthur 70

D

Dark Sky Station 5, 15-16, 29-33, 49, 63, 65,
68-69, 71, 81-82, 85-86, 90-92, 94-99, 103,
109, 115, 117, 122, 124-125, 131, 133, 138-
139, 141, 145, 173, 179-181, 183, 195-196,
198
Darth Vader 68
DC-X 10

Deep Space One 89, 115
Deutsch de la Meurthe prize 53, 76
Dnepr rocket 62Adams, Douglas 69
Dragonlady 47
DSS 13, 15, 29-32, 68, 85, 88, 94-95, 98-103,
130, 137-138, 140, 151, 179-183
Dumont airship 53, 76

E

e.coli 17, 46
ECHO 27, 59-60, 69
Elves 17, 42
Exobase 41
Explorer IX 41, 59

F

FAA 147
Fabrication of Inflatable Re-entry Structure
for Test 57
Farside 55
Fuller, Buckminster 85, 87

G

Gagarin, Yuri 10
Genesis I 62
Gerlach, Nevada 151
Giffard, Henri 53
Glenn, John 10
Gnomes 17, 43, 147
Goodyear 13, 15, 23, 76, 113
Gossamer 14-16, 48, 75
Graf Zeppelin 15, 23, 29, 76
Gudenoff, Constantin 54
Gulliver's Travels 67
Gusmao, Bartolomeu Lourenco de 52

H

Halley's comet 60
Helios 48
Helium bulge 41, 59
Hindenburg 19, 76
Hyderabad 44

I

IMP 56-57, 62, 121
Inflato-Plane 113
Ionosphere 41
IRDT 62

Accompanying DVD includes:

Remarkable footage taken during various trips to space
aboard JP Aerospace's experimental test program.

Texas Launch	Flying with the Sons of the San Joaquin
The Ring	Inside with Space Vacuum
Rockets	Scale Tandem Test
There and Back	New Airship Propellers
Ascender 90	JPA Promo
Ascender 90 at Night	Prehistoric JPA
Platforms	Away 33
Weird Balloon	Away 34
Rocket from Below	The Sky
Away 27 Hop	Countdown
Dark Sky Station 2	